CIRIA C648

London, 2006

Control of water pollution from linear construction projects

Technical guidance

E Murnane	HyderConsulting
A Heap	HyderConsulting
A Swain	Edmund Nuttall

CIRIA sharing knowledge ■ building best practice

Classic House, 174-180 Old Street, London EC1V 9BP, UK
TEL: +44 (0)20 7549 3300   FAX +44 (0)20 7253 0523
EMAIL enquiries@ciria.org
WEBSITE www.ciria.org

Summary

This publication provides guidance to clients, consultant, designers, contractors and regulators on how to plan and manage water pollution from road, railway, pipeline, waterway and other linear construction projects. It is divided into three sections:

- **Part A Characteristics of linear projects and understanding water pollution** provides an introduction to water pollution and looks at the types and characteristics of water environments.

- **Part B Planning and design** is concerned with the design, planning and programming of a project and the measures that can be taken at these critical stages to minimise water pollution during construction.

- **Part C Construction** provides guidance on the construction phase of a project and covers the management and control of water onsite and pollution prevention measures for key construction activities.

This guidance is fully cross-referenced and illustrated and is intended to be a user-friendly reference guide to support previous CIRIA publications on the subject.

Control of water pollution from linear construction projects. Technical guidance

Murnane, E; Heap, A; Swain, A

CIRIA

CIRIA C648   © CIRIA 2006   RP708

ISBN-13: 978-0-86017-648-0
ISBN-10: 0-86017-648-7

British Library Cataloguing in Publication Data

A catalogue record is available for this book from the British Library.

Keywords
Construction management, environmental good practice, pollution prevention, rivers and waterways, site management, water quality

Reader interest	Classification	
Pollution prevention, linear construction, surface water, groundwater, construction cycle, pollution migration	AVAILABILITY	Unrestricted
	CONTENT	Advice/guidance
	STATUS	Committee-guided
	USER	Construction professionals and managers

Published by CIRIA, Classic House, 174–180 Old Street, London EC1V 9BP, UK.

All rights reserved. No part of this publication may be reproduced or transmitted in any form or by any means, including photocopying and recording, without the written permission of the copyright-holder, application for which should be addressed to the publisher. Such written permission must also be obtained before any part of this publication is stored in a retrieval system of any nature.

This publication is designed to provide accurate and authoritative information in regard to the subject matter covered. It is sold and/or distributed with the understanding that neither the authors nor the publisher is thereby engaged in rendering a specific legal or any other professional service. While every effort has been made to ensure the accuracy and completeness of the publication, no warranty or fitness is provided or implied, and the authors and publisher shall have neither liability nor responsibility to any person or entity with respect to any loss or damage arising from its use.

Acknowledgements

This publication was produced as a result of CIRIA Research Project 708, "Control of water pollution from linear construction projects" and was written by Ms Emma Murnane and Mr Andy Heap of Hyder Consulting Ltd and Mr Andrew Swain of Edmund Nuttall Ltd. Additional contributors were Mr Luke Stalley, Mr Bob Sargent, Dr Amy Davis, Mr Simon Witney, Ms Abigail Frost, Ms Sarah Hammond and Mr Daniel Palmer of Hyder Consulting, and Ms Maria Jarosz of Edmund Nuttall Ltd.

Authors

Emma Murnane specialises in environmental management and co-ordination of construction projects including major road and pipeline schemes in urban and rural settings, as well as advising on sustainable construction. She is also author of CIRIA SP156 *Control of water pollution from construction sites – guide to good practice*.

Andy Heap is a project manager for major civil engineering projects (road, rail, transmission lines and pipelines) and specialises in environmental impact assessments and construction environmental management. He is also author of CIRIA C532 *Control of water pollution from construction sites. Guidance for consultants and contractors*, and managed the preparation of CIRIA SP156 *Control of water pollution from construction sites – guide to good practice*.

Andrew Swain is environment manager for Edmund Nuttall Ltd; he specialises in providing in-house environmental advice and technical support to project teams involved in a wide variety of major civil engineering schemes, including linear construction projects.

Steering group

Following CIRIA's tradition of collaboration, the study was guided by a steering group of individuals involved in, or with an interest in the control of water pollution from linear construction sites and related risk mitigation. CIRIA would like to express its thanks and appreciation to all members of the project steering group for their commitment and valued comments throughout the project.

Chairman	Dr Nick O'Riordan	Arup
Attending members	Dr Paul Beckwith	British Waterways
	Mr Barry Beecroft	The BOC Foundation
	Mr Martin Brock	Balfour Beatty
	Mr Phil Chatfield	Environment Agency
	Mr Ian Clarke	Morgan Est
	Dr Mike de Silva	Transport for London
	Mr Peter Fisher	Costain Ltd
	Mr Dave Gibson	Alfred McAlpine
	Mr Sam Hall	Carillion
	Ms Sarah Hides	Defra
	Mr Gareth Jones	MJ Gleeson Group plc
	Mr Howell Jones	Amec Group Ltd.
	Ms Liz McDonnell	Defra
	Mr Peter Martin	Black & Veatch
	Mr Grahame Newman	British Waterways

Mr Simon Price	Highways Agency
Mr Stan Redfearn	The BOC Foundation
Mr Will Rogers	URS Corporation
Mr S Santhalingam	Highways Agency
Dr Steve Yeoman	National Grid Transco

Corresponding members

Mr John Lonergan	Balfour Beatty
Mr Alistair McNeill	Scottish Environment Protection Agency
Ms Elizabeth Morrison	Scottish Executive

CIRIA project management

The project was initially developed by Ms **Marianne Scott** and subsequently managed by Dr **Das Mootanah** and Ms **Victoria Cole**.

Funders

The project was funded by The BOC Foundation, the Highways Agency, Defra, the Environment Agency and the Scottish Executive.

In addition to the steering group members, CIRIA is grateful to the following organisations for providing photographs, supporting information, case studies and feedback on this study:

ADAS
Amstar
Black & Veatch Costain
Crane Consultants
Dr Peter Howsam, Cranfield University
DTI
Dwr Cymru Welsh Water
Entrepose UK
Ewan Associates Ltd
Faithful & Gould
Morrison Construction
Mowlem Civil Engineering
Network Rail
WJ Groundwater Ltd.

> This guidance provides best practice advice and is intended to supplement, rather than replace, any contractual requirements, consultation with regulators or company procedures.

Scope

In this guidance, the following types of linear projects are considered:

- roads – including new motorways, dual carriageways, bypasses, road widening, tunnels, bridges
- railways – including all new railway lines and on-line infrastructure track upgrades, tunnels, bridges, light rail systems, tramways
- pipelines – including new and replacement water, sewerage, oil, gas and chemical pipelines
- cables – including high- and low-voltage electricity supplies below ground or overhead, telecommunications cables
- watercourses – including canals, flood defences, river diversions.

This guidance is specifically aimed at linear projects, although much of the guidance is applicable to any development project. It addresses the control of water pollution throughout the whole project cycle, from the design of a scheme through to construction and commissioning.

Small urban projects such as minor utility and road works, pathways and urban electric cables are not included, although the guidance would be of relevance and interest. Excluded from the scope are coastal and offshore works. CIRIA has published other guidance on managing environmental issues in construction that may be of use to these types of project (see "Other CIRIA guidance", below).

Also excluded from the scope is the maintenance and operational phase of projects (except insofar as any permanent works may be used for temporary works during construction) and decommissioning. The guidance does, however, address issues encountered on upgrades and on-line replacement.

The guidance sets out generic best practice and procedures for controlling water pollution from construction sites in England, Wales, Scotland and Northern Ireland. The reader should note that there are regional legislative and regulatory variations; attention is drawn to these where relevant. However, anyone intending to implement the practices and procedures set out in this guidance should ensure the work complies with the relevant regional variations.

The term "the environmental regulator" refers to:

- Environment Agency (EA) with jurisdiction in England and Wales
- Scottish Environment Protection Agency (SEPA) with jurisdiction in Scotland
- Environment and Heritage Service (EHS) with jurisdiction in Northern Ireland.

The term "the conservation bodies" refers to:

- Natural England (formerly English Nature, the Countryside Agency and the Rural Development Service) with jurisdiction in England
- Countryside Council for Wales (CCW) with jurisdiction in Wales
- Scottish Natural Heritage (SNH) with jurisdiction in Scotland
- Environment and Heritage Service (EHS) with jurisdiction in Northern Ireland.

Target readership

All personnel involved in the promotion, design, construction and maintenance of infrastructure developments have to be aware of their environmental obligations and the benefits that best practice will bring to all stages of a construction project. Decisions taken at the planning and design stage of a project can have a significant impact on the control of water pollution once the project reaches the construction stage. This guidance is of particular relevance to those working on linear projects (roads, railways, pipelines etc), although it can also be applied to most construction sites. The reader should also refer to "Other CIRIA guidance", below for publications about other types of construction site (eg coastal).

This guidance is written for a wide range of readers including:

- clients/promoters
- designers
- environmental consultants
- construction project managers
- senior site engineers and site agents
- site environmental managers
- regulators.

This book is of relevance to all construction personnel. It is supplemented by a *Site guide* (CIRIA publication C649), which is aimed particularly at the following:

- site engineers and construction managers
- site foremen and site supervisors.

How to use this book

This book is arranged in three parts to allow easy reference at different stages of a project:

Part A – Characteristics of linear projects and understanding water pollution

Part B – Planning and design

Part C – Construction.

Part A provides an introduction to the unique characteristics of linear projects, explains what is meant by "water pollution" and provides information on the types and characteristics of water environments. Chapter 2 in particular provides useful information to help understand the water environment (surface water and groundwater).

Reducing the risk of water pollution from a construction site starts well before the construction stage. Part B is concerned with the design, planning and programming of a project. At this stage critical decisions are made in terms of route selection and scheme design, which can help to minimise the risk of water pollution throughout the project.

Part C provides guidance on the construction phase of a project. It applies to projects for which any necessary planning approval has been obtained, to contractors requiring guidance at tender or any later stage of the construction phase and to clients and regulators in supervising construction works. This section covers the management and control of water on site and pollution prevention measures for key construction activities.

Developing a project is an iterative process, and although this guidance is divided into planning/design and construction phases, those working in each phase need to be aware of the issues in the subsequent or preceding phase.

This document is accompanied by a *Site guide* (CIRIA C649), which provides key guidance for use on site and can also be read as a standalone document.

Throughout this *Technical guide* and the *Site guide*, the following symbols are used to identify the types of information being provided:

| Plan ahead | Take note | Checklist | Information |

| The law | Case study | Key guidance |

Other CIRIA guidance

This document is one of three published by CIRIA that provide key guidance on controlling water pollution from construction:

- C532 *Control of water pollution from construction sites: guidance for consultants and contractors* (Masters-Williams *et al*, 2001)
- SP156 *Control of water pollution from construction sites – guide to good practice* (Murnane *et al*, 2002), comprising training presentation, site inspection checklists, best practice guidance sheets, toolbox talks and a poster.

Previously published guidance on environmental issues in construction from CIRIA includes:

- C533 *Environmental management in construction* (Uren and Griffiths, 2000)
- C584 *Coastal and marine environmental site guide* (Budd *et al*, 2003)
- C587 *Working with wildlife. A resource and training pack for the construction industry* (Newton *et al*, 2004a)
- C613 *Working with wildlife pocket book* (Newton *et al*, 2004b)
- C650 *Environmental good practice on site (second edition)* (Chant-Hall *et al*, 2005a)
- C651 *Environmental good practice on site – pocket book* (Chant-Hall *et al*, 2005b)
- SP120 *A client's guide to greener construction* (CIRIA, 1995)
- SP141V *Building a cleaner future* (CIRIA, 1996), joint CIRIA and Environment Agency training pack, including video, booklet and poster.

Other related CIRIA guidance:

- C643 *The potential for water pollution from railways* (Osborne and Montague, 2005)
- SP125 *Control of risk – a guide to the systematic management of risk from construction* (Godfrey, 1996).

Contents

Summary 2
Acknowledgements 3
Scope 5
Target readership 6
How to use this book 7
Other CIRIA guidance 8
List of figures 13
List of tables 15
Glossary 16
Abbreviations 24

Part A CHARACTERISTICS OF LINEAR PROJECTS AND UNDERSTANDING WATER POLLUTION 27

1 Introduction 29
    1.1 Characteristics of linear projects 29

2 Water environments 31
    2.1 Surface water 32
    2.2 Groundwater 33
    2.3 Water on site 36

3 Water pollution and the law 37
    3.1 Types and sources of pollution 37
    3.2 Pollution offences 38
    3.3 Water Framework Directive 40
    3.4 Responsibility for and costs of pollution 41

Part B PLANNING AND DESIGN 43

4 Introduction 45

5 Scheme design and land take 47
    5.1 Route selection 47
    5.2 Design 48
    5.3 Land take 49

6 Stakeholder consultation 51

7 Development consent 55
    7.1 Introduction 55
    7.2 Enabling and non-planning legislation 55
    7.3 Permitted development 56
    7.4 Planning permission 57
    7.5 Environmental impact assessment 58

8	Site investigations and monitoring	59
	8.1 Introduction	59
	8.2 Baseline monitoring	59
	8.3 Site investigation data used to manage pollution risk	60
	8.4 Pollution caused by site investigations	61
9	Programming and seasonal influences	63
10	Contracts	65
	10.1 Type of contract	65
	10.2 Tender and contract specification	66
	10.3 Liability	69

Part C CONSTRUCTION 71

11	Introduction	75
12	Site planning	77
	12.1 Introduction	77
	12.2 Environmental management plans	77
	12.3 Risk assessment and control	78
	12.4 Consultation with regulators and other organisations	79
	12.5 Programming and seasonal pollution control issues	80
	12.6 Roles and responsibilities	83
	12.7 Emergency procedures	85
13	Licences and consents	87
	13.1 Discharging to sewer	87
	13.2 Discharging to surface water or groundwater	88
	13.3 Abstracting and dewatering	89
	13.4 Works in or near water	90
	13.5 Works in tidal waters	92
14	Monitoring	93
	14.1 Legal requirements	93
	14.2 Benefits of monitoring	93
	14.3 What to monitor	94
	14.4 When to monitor	94
	14.5 How to monitor	95
	14.6 Records	98
15	Emergency and contingency planning	101
	15.1 Risk assessment	101
	15.2 Emergency plans and procedures	102
	15.3 Training and testing	103
	15.4 Equipment	104
	15.5 Corrective action	105

16	Site set-up	107
	16.1 Introduction	107
	16.2 Site drainage and water features	108
	16.3 Water supply	110
	16.4 Water use	110
	16.5 Wastewater disposal	112
	16.6 Storage and use of materials	113
	16.7 Waste management	116
	16.8 Fuel and oil	117
	16.9 Site security	122
17	Adjacent land and water use	125
	17.1 Protecting adjacent land and water uses	125
	17.2 Protecting the site from adjacent activities	127
	17.3 Additional land take	128
18	Runoff and sediment control	129
	18.1 Introduction	129
	18.2 Preparing an erosion and sediment control plan	130
	18.3 Estimating runoff	132
	18.4 Flooding	137
	18.5 Estimating sediment generation	138
	18.6 Erosion and sediment control measures	140
	18.7 Protecting existing and pre-construction drainage	151
	18.8 Sustainable drainage systems (SUDS)	154
19	Water treatment methods and disposal	155
	19.1 Introduction	155
	19.2 Sediment	155
	19.3 Concrete and cementitious material	167
	19.4 Fuel and oil	167
	19.5 Metals	169
	19.6 Ammonia and oxygen levels	170
	19.7 Sewage	170
	19.8 Disposal options and temporary outfalls	170
20	Works in or near water	175
	20.1 Planning the works – legal requirements	175
	20.2 Pollution controls	176
	20.3 Access and haul routes across water	179
	20.4 Trenchless construction	181
	20.5 Open excavations and diversions	184
	20.6 Overpumping	186
	20.7 Bank works	187
	20.8 Works near watercourses	188
	20.9 Works in the floodplain	190
	20.10 Works over water	190

21	Excavations and dewatering	193
	21.1 Legal requirements	193
	21.2 External dewatering (groundwater)	193
	21.3 Internal dewatering (excavations)	194

22	Concrete and grouting activities	199
	22.1 Legal requirements	199
	22.2 Alternative methods	199
	22.3 On-site batching	200
	22.4 Transport and placement	200
	22.5 Tunnelling, thrust-boring and pipejacking	202

23	Contaminated land	205
	23.1 Introduction	205
	23.2 Investigation and assessment	205
	23.3 Development of specific mitigation	205
	23.4 Managing unexpected contamination	206

24	Ecology	209
	24.1 Legal protection	209
	24.2 Construction impacts	211
	24.3 Vegetation clearance and landscaping	212

APPENDICES 213

A1	EIA legislation in relation to linear projects	215
A2	Calculating site runoff rates	217
A3	Guidance on the optimal timing for carrying out ecological surveys and mitigation	221
A4	Internationally, nationally and locally designated sites	223
	References and bibliography	225
	Legislation	231
	Standards	232
	EA publications	233
	Websites	234

Figures

Figure 2.1	The hydrologic cycle	31
Figure 4.1	Stages in the development of project proposals	46
Figure 11.1	Construction activities and water pollution issues discussed in this guidance	76
Figure 12.1	ISO 14001 elements	77
Figure 12.2	Rainfall amount (mm) annual average 1971–2000	80
Figure 14.1	Example proforma for visual inspections of river	100
Figure 15.1	Example emergency response procedure	102
Figure 16.1	Mobile bunded bowser	118
Figure 16.2	Integrally bunded tank	119
Figure 16.3	Open bunded tank	119
Figure 16.4	Storage site guidelines	120
Figure 17.1	Silt pollution at fish farm	126
Figure 17.2	Silt pollution in river	126
Figure 17.3	Construction works	126
Figure 17.4	Improved working area	126
Figure 18.1	Rainfall amount annual average (mm) 1971–2000	135
Figure 18.2	Relationship between stream flow velocity and particle erosion, transport and deposition	138
Figure 18.3	Before and after seeding: erosion gulleys on non-vegetated slope	141
Figure 18.4	Tracking and furrowing	142
Figure 18.5	Coconut matting and silt fence	143
Figure 18.6	Haul road	144
Figure 18.7	Haul route runoff	144
Figure 18.8	Stabilised construction entrance	145
Figure 18.9	Diversion drain alongside sloping pipeline easement	146
Figure 18.10	Diversion drain (lined with geotextile)	146
Figure 18.11	Level spreader	147
Figure 18.12	Slope drain	147
Figure 18.13	Slope drain detail	148
Figure 18.14	Check dams	148
Figure 18.15	Fabric silt fence at toe of stockpile	149
Figure 18.16	Straw bale installation	150
Figure 18.17	Straw bale (or hay bale) and geotextile fence installation	151
Figure 18.18	Silt fence and straw bale arrangement to control silty runoff	151
Figure 18.19	Kerb inlet control	152
Figure 18.20	Storm drain control	152
Figure 18.21	Land drain uncovered during construction work	153
Figure 18.22	Land drains often blend into the environment, so may be hard to identify	153
Figure 18.23	Silty water from a land drain entering a watercourse	153

Figure 19.1	Silty water discharging through land drain into ditch	156
Figure 19.2	Unlined excavated pond adjacent to working easement	160
Figure 19.3	Series of lagoons	160
Figure 19.4	New unlined bunded lagoon	160
Figure 19.5	Skips used as settlement tanks	161
Figure 19.6	Settlement pond with straw bale filter and oil boom	169
Figure 19.7	Baffles on discharge hoses	172
Figure 19.8	Bank protection for discharge outfall	172
Figure 20.1	Straw bales filtering silt in newly constructed stream	172
Figure 20.2	Silt mat	177
Figure 20.3	Correct installation of oil boom	179
Figure 20.4	Culverted haul route crossing (note straw bales and "terram" to filter silty runoff)	180
Figure 20.5	Haul route culverts sized for maximum flood flow	181
Figure 20.6	Sand bags along existing bridge protecting river below from haul road runoff	181
Figures 20.7	First grout breakout controlled with sand bags	183
Figure 20.8	Second breakout controlled with sand bags in stream	183
Figure 20.9	Grout emerging from borehole into lagoon	183
Figure 20.10	Clean discharge from lagoon entering stream through straw bales	183
Figure 20.11	Flume pipes carrying water flow during open-cut crossing of stream for pipeline installation	184
Figure 20.12	Silt trap comprising bags of sand used during watercourse diversion	185
Figure 20.13	River diversion to allow open-cut crossing. Lined with geotextile to avoid silt generation	185
Figure 20.14	Large stones used for bed protection at pipe outfall	187
Figure 20.15	Reinstated open-cut river crossing, with no hard protection, to allow revegetation	188
Figure 20.16	Timber boards and coir matting installed as river bank stabilisation	188
Figure 20.17	Sheeting of Pontcysyllte Aqueduct over River Dee SSSI	190
Figure 20.18	Beams for the M1A1 bridge being erected over the River Aire	191
Figure 21.1	Discharging water from pumping activities has caused silt pollution in the stream	195
Figure 21.2	Permit to pump	196
Figure 21.3	Use of a ladder to keep pump off the base of the excavation to avoid disturbing sediments	197
Figure 21.4	Use of an excavator arm for the same purpose	197
Figure 24.1	Ecological survey of stream in SSSI before construction of a pipeline crossing	211
Figure A3.1	Wildlife year planner	221

Tables

Table 2.1	Construction activities that pose a high risk of surface water impact	32
Table 2.2	Construction activities that pose a high risk of groundwater impact	35
Table 2.3	Definitions of water and wastewater	36
Table 3.1	Pollution types and sources	38
Table 5.1	Examples of design controls to avoid water pollution hazards	48
Table 6.1	Consultees with interests in protecting the water environment	52
Table 12.1	Rate of salt application	82
Table 14.1	Example water monitoring requirements	94
Table 15.1	Example measures to reduce pollution incidents	101
Table 15.2	Emergency equipment	104
Table 16.1	Advantages and disadvantages of biodegradable oils	121
Table 17.1	Sensitive receptors	125
Table 17.2	Pollution prevention measures – summary and further guidance	127
Table 17.3	Protection from off-site pollution – summary and further guidance	128
Table 18.1	Typical infiltration rates for various soils	133
Table 18.2	Mean annual flood peak flow for catchments <50 ha	134
Table 18.3	Soil classes	134
Table 18.4	Factors for different return periods	134
Table 18.5	Regional factors for scaling mean annual flood	136
Table 18.6	Range of control for silt fences	149
Table 19.1	Typical infiltration rates for various soils	156
Table 19.2	Theoretical range of retention times for a variety of particle sizes	162
Table 23.1	Construction activities with pollution risks from land contamination	206
Table A2.1	Mean annual flood peak flow for catchments < 50 ha	217

Glossary

Abstraction	Removal of water from surface water or groundwater, usually by pumping.
Abstraction licence	Permission to abstract surface water or groundwater, subject to conditions laid down in the licence, issued by the relevant environmental regulator.
Algae	Simple plants ranging from single cells to large plants.
Ammonia	A water-soluble chemical compound, produced by the decomposition of organic material. Ammonia affects the quality of fisheries and the suitability of abstractions for potable water supply. Used as a water quality indicator. Ammonia is a List II substance (see below).
Aquifer	A source of **groundwater** comprising water-bearing rock, sand or gravel capable of yielding significant quantities of water (see also **unconfined aquifer**).
Artificial drainage system	A constructed drainage system such as a drain, sewer or ditch.
Balancing pond	Pond or lagoon, also known as a retention pond, where the inflow varies and the outflow is restricted to a maximum rate to contain water permanently. Often used to control runoff and to prevent flooding by "balancing" the flow – see also **settlement pond**, **detention pond**.
Bentonite	A colloidal clay, largely made up of the mineral sodium montmorillonite; typically used as a lubricant for drilling and pipejacking.
Biodegradable	Capable of being decomposed by bacteria or other living organisms.
Biochemical oxygen demand	BOD is the measure of the concentration of biodegradable organic carbon compounds in solution. Used as a water quality indicator.
Brownfield site	A site that has been previously used or developed and is not currently fully in use, although it may be partially occupied or utilised (see also **greenfield site**).
Bund	A barrier, dam or mound used to contain or exclude water (or other liquids). Can either refer to a bund made from earthworks material, sand etc or a metal/concrete structure surrounding, for example, a fuel tank.
Caisson	A cylindrical or rectangular ring wall usually formed from pre-cast concrete segments and used for excluding water or supporting soft ground in deep excavations.
Catchment	The area from which water or **runoff** drains to a specified point (eg to a reservoir, river, lake, borehole).
Chemical oxygen demand	COD is the measure of the amount of oxygen take up by chemical oxidation of a substance in solution. Used as a water quality indicator.

Cofferdam	A temporary dam, usually of sheet piling driven into the ground to exclude water, provide a vertical face support to an excavation and/or provide access to an area that is otherwise submerged or waterlogged.
Conceptual model	A textural or graphical simplified representation of the behaviour of a natural system, including identification of the relationships between contamination source(s), pathway(s) and receptors.
Cone of depression	A depression in the groundwater table shaped like an inverted cone that develops around a well from which water is being withdrawn.
Construction cycle	The sequence of events or activities carried out in the development of a construction project.
Construction (Design and Management) Regulations	The CDM Regulations emphasise the importance of addressing health and safety issues at the design phase of a construction project.
Contaminant	See **source**.
Contaminated land	Non-legislative definition: land with elevated concentrations of hazardous substances that either occur naturally, or, more often, result from past or current industrial activities, and present an actual or potential hazard to potential **targets**.
Controlled waters	Defined by the Environment Act 1995, they refer to almost all natural water bodies in England, Scotland and Wales, including all rivers, streams, lochs, ditches, canals, burns, ponds and groundwater. Termed waterways in Northern Ireland.
Croy	A construction in a river channel to create bed scour and form a holding area for fish. If constructed on a scale disproportionate to the river channel, a croy can seriously affect the integrity and stability of the river system.
Culvert	Covered channel or pipe that forms a watercourse below ground level.
Cyprinid fishery	Waters in which coarse fish (such as pike, perch and eel) are found.
Desk study	Desk-based research and interpretation of historical, archival and current information to understand the environmental setting and sensitivity of a site, establish previous land uses and potential contamination, and identify potential pollution linkage scenarios.
Detention pond/tank	A pond or tank that holds water temporarily, often to prevent flooding, which is dry under normal conditions (in contrast to a retention pond, which holds water permanently). See also **balancing pond, settlement pond**.
Dewatering	The removal of groundwater/surface water to lower the water table or to empty an area, such as an excavation, of water.

Diffuse pollution	Pollution that arises from sources other than **point sources**. Usually refers to surface water runoff, road runoff or leaching of land into groundwater.
Discharge consent	Permission to discharge effluent, subject to conditions laid down in the consent, issued by the relevant environmental regulator.
Dissolved oxygen (DO)	The amount of oxygen dissolved in water. Oxygen is vital for aquatic life, so this measurement is a test of the health of a river. Used as a water quality indicator.
Drawdown	The distance between the static water level and the surface of lowered water level.
Dry weather flow	Rate of flow in a watercourse, drain or sewer in dry weather conditions; also known as baseflow. In some circumstances referred to as Q95 flow (flow that is exceeded 95 per cent of the time).
Dust	Airborne solid matter up to about 2 mm in size diameter.
Duty of care	The implication of the duty of care is that toxic materials are monitored and administered by an appropriate system each time they pass from one individual to another or from one process to another. Important information regarding the nature of the material and any appropriate emergency action should also be passed on.
Ecology	All living things, such as trees, flowering plants, insects, birds and mammals, and the habitats in which they live.
Ecosystem	Interdependence of species in the living world with one another and their non-living environment.
Environment	Both the natural environment (air, land, water resources, plant and animal life) and the habitats in which they live.
Environmental impact assessment	A technique used for identifying the environmental effects of development projects. As a result of European Directive 85/337/EEC (as amended 1997), this is a legislative procedure to be applied to the assessment of the environmental effects of certain public and private projects that are likely to have significant effects on the environment.
Environmental regulators	These include the Environment Agency (in England and Wales), the Scottish Environment Protection Agency and the Environment and Heritage Service in Northern Ireland.
Environmental statement	A document, or series of documents, assessing the environment effects of a development produced as a result of an **environmental impact assessment** and usually submitted in support of a planning application.
Estuary	A semi-enclosed body of water in which seawater is substantially diluted with freshwater entering from river and land drainage.
Ex situ	Located away from the place of origin or place of work. Can refer to excavated materials (eg *ex-situ* remediation) or processes taking place off site. See also ***in situ***.

Fauna	The animals found in a particular physical environment.
Field drain	System of piped or gravel drains to control the water table in agricultural land (also **land drain**).
Filter strip	Vegetated area of land used to accept surface runoff as sheet flow from an upstream area.
Floc	A small, loosely held mass or aggregate of fine particles formed in a fluid through precipitation or aggregation of suspended particles.
Floodplain	Area of land that borders a watercourse, an estuary or the sea, over which water flows in time of flood, or would flow but for the presence of flood defences where they exist.
Flora	The plants found in a particular physical environment.
French drains	Shallow trench filled with fine aggregate (gravel) constructed at a gradient to allow surface and sub-surface water to drain away.
Greenfield site	Any land that remains undeveloped (see also **brownfield site**).
Grip	A small channel or ditch cut into the ground on the uphill side of an excavation to channel rainwater away from the excavation.
Groundwater	Defined by the EC Groundwater Directive (80/68/EEC) as "all water which is below the surface of the ground in the saturation zone and in direct contact with the ground or subsoil".
Groundwater protection zone	These zones provide an indication of risk to groundwater supplies, for which source protection zones have been identified, that may result from potential pollution activities and accidental release of pollutants. Generally the closer the activity or release to a groundwater source the greater the risk. Three zones (inner, outer and total catchment) are usually defined.
Grout	A fluid mixture of cement and water of such a consistency that it can be forced through a pipe and placed as required. Various additives – eg sand, **bentonite** and hydrated lime – may be included in the mixture to meet certain requirements.
Grouting	The operation by which grout is placed.
Gulley	An opening in which rain or wastewater is collected before entering a drain.
Gully erosion	The erosion of soils by surface runoff, resulting typically in steep-sided channels and small ravines (gullies).
Hazard	An activity, situation or substance with the potential to cause adverse effects.
Heavy metal	Loosely, metals with a high atomic mass, often used in discussion of metal toxicity. No definitive list of heavy metals exists, but they generally include cadmium, zinc, mercury, chromium, lead, nickel, thallium, silver. Some

	metalloids, eg arsenic and antimony, are classified as heavy metals for discussion of their toxicity.
In situ	Located at the place of origin or place of work. Can refer to materials in the ground (eg *in-situ* remediation) or processes taking place on site. See also *ex situ*.
Internal drainage board (IDB)	Body with powers and duties relating to **ordinary watercourses** within defined local drainage areas in England and Wales.
Invertebrates	Animals that lack a vertebral column. This includes many groups of animals used for biological grading, such as insects, crustaceans, worms and molluscs.
Land drain	Drain used in agriculture (see **field drain**) to control the water table and reduce the frequency with which land becomes waterlogged.
Leaching	The process during which soluble materials may be removed from the soil by water percolating through it.
List I substance	A controlled substance as defined by the Groundwater Regulations 1998 and the Dangerous Substances Directive (76/464/EEC). **List I substances** are considered the most dangerous in terms of toxicity, bioaccumulation and persistence. These controls prevent their discharge to the environment.
List II substance	A controlled substance as defined by the Groundwater Regulations 1998 and the Dangerous Substances Directive (76/464/EEC). They are less toxic than List I substances but are still capable of harm, hence their discharge to the environment is limited.
Local planning authority	Body responsible for planning and controlling development, through the planning system
Main river	A watercourse identified by Defra, the Welsh Assembly Government and the environmental regulator and designated on a statutory map of main rivers.
Microtunnelling	Method of steerable remote control **pipejacking** to install pipes of internal diameter less than that permissible for man-entry.
Nature conservation body	The organisations that have regional responsibility for promoting the conservation of wildlife and natural features: Countryside Council for Wales, English Nature, Northern Ireland Environment and Heritage Service, and Scottish Natural Heritage.
Nutrient	A substance providing nourishment for living organisms (such as nitrogen and phosphorous).
Open cut	Traditional method of installation with pipes laid in an open trench, which is subsequently backfilled.
Ordinary watercourse	A watercourse that is not a private drain and is not designated a **main river**.
Outfall	End of a temporary or permanent pipeline from which water (or other effluent) is discharged. Can refer either to the end of a length of pipe or to a dedicated structure.

Pathway	The route by which potential contaminants may reach targets.
Piezometer	An instrument installed to allow groundwater level measurement and sampling.
Permeability	The property or capacity of a rock, sediment or soil for transmitting a fluid.
Pipejacking	Method for directly installing pipes by hydraulic or other jacking from a driveshaft such that the pipes form a continuous string in the ground.
Point-source pollution	Pollution that arises from point discharges, which are at definable locations, such as an outfall or discharge pipe. See also **diffuse pollution**.
Pollution	The introduction of a substance or energy that has the potential to cause harm to the environment. Pollutants can include silty water, oils, chemicals, litter, mud, noise and heat.
Pollution linkage	The **source-pathway-receptor** relationship which results in, or may potentially result in the receptor being impacted by the source.
Polymer	A chemical formed by the union of many monomers (a molecule of low molecular weight). Polymers are used with other chemical coagulants to aid in binding small suspended particles to form larger chemical flocs for easier removal from water.
Receptor	The entity or **target** (eg surface water, groundwater, wetland, human, animal, building etc) that is vulnerable to the adverse effects of a hazardous substance, pollution or contaminant.
Recharge	The addition of water to the groundwater system by natural or artificial processes.
Recycling	Reusing water that has been previously used in an initial activity and subsequently processed by filtration, settlement or other means to remove pollutants.
Reed bed	Area of grass-like marsh plants. Artificially constructed reed beds can be used to accumulate suspended particles and associated heavy metals or to treat small quantities of partially treated sewage effluent.
Return period	The average interval between events of a given magnitude (mm for rainfall events). Also referred to as a percentage probability, ie a 50-year return period event has a 2 per cent probability of occurring in any one year.
Reuse	Putting objects back into use, without processing, so that they do not remain in the waste stream.
Risk	The likelihood that a **hazard** will actually cause its adverse effects, together with a measure of the effect.
Risk assessment	A process requiring an evaluation of the risk that may arise from identified hazards, combining the factors contributing to the risk and then estimating their significance.

Runoff	The water from rain, snowmelt or irrigation that flows over the land surface and is not absorbed into the ground, but which instead flows into streams or other surface waters or land depressions.
Salmonid fishery	Waters in which game fish (such as salmon, trout, grayling and whitefish) are found.
Sedimentation tank/pond	A pond or tank used to reduce turbidity of water allowing solid particles to settle out (see **suspended solids**). The inflow and outflow are equal. See also **storage pond**, **detention pond**, **settlement pond**.
Sediment	Particles such as soil, sand, clay, silt and mud, which comprise the main water pollutant from construction. See **suspended solids**.
Settlement tank/pond	Pond, tank or lagoon used to hold water in order to reduce turbulence thus allowing solid particles to settle out. See also **balancing pond**, **storage pond** and **detention pond**.
Sewer	Pipeline or other construction, usually underground, designed to carry wastewater and/or surface water from more than one source.
Sewerage	See **sewer system**.
Sewer system	Network of pipelines and ancillary works that conveys wastewater and/or surface water from drains to a treatment works or other place of disposal.
Silt	The generic term for particles with a grain size of 4–63 mm, ie between clay and sand.
Site of special scientific interest (SSSI)	An area of land or water notified under the Wildlife and Countryside Act 1981 (as amended) as being of geological or nature conservation importance, in the opinion of Countryside Council for Wales, English Nature or Scottish Natural Heritage.
Source	The activity or process producing a hazardous substance or contaminant that may adversely impact a **receptor** via a **pathway**.
Special area of conservation (SAC)	Established under the EC Habitats Directive (92/43/EEC), implemented in the UK by the Conservation (Natural Habitats, &c) Regulations 1994, and the Conservation (Natural Habitats, etc) Regulations (Northern Ireland) 1995. The sites are significant in habitat type and species and are considered in greatest need of conservation at a European level. All UK SACs are based on SSSIs, but may cover several separate but related sites.
Storage pond/tank	Pond (sometimes called a lagoon) or tank used to hold water, with no outflow.
Strata	Distinct layers of soil or rock.
Sump	A hole or pit that may be lined or unlined and is used to collect water to enable pumping out.

Surface water	Water that appears on the land surface that has not seeped into the ground, ie lakes, river, streams, standing water, ponds, precipitation. See also **runoff**.
Surface water sewer	Sewer system carrying surface water runoff (rainfall) from hardstanding, roads and building roofs etc.
Sustainable drainage systems (SUDS)	A sequence of management practices and control structures designed to drain surface water in a more sustainable manner than some conventional techniques. Typically, SUDS are used to attenuate rates of runoff from development sites and can also have water purification benefits.
Suspended solids	General term describing particles such as sand, clay, silt, mud in suspension in water. Used as a water quality indicator.
Swale	An open grassed drainage channel in which surface water may be conveyed and allowed to infiltrate and which can remove some pollutants.
Target	See **receptor**.
Trade effluent	Wastewater discharge resulting wholly or in part from any industrial or commercial activity, including construction. The only effluents that are not classed as trade effluent are clean, uncontaminated surface water (ie clean rainwater that has not been contaminated) and domestic sewage.
Trenchless technology	Methods for utility and other line installation, replacement, repair and inspection, with minimum excavation from the ground surface (see **pipejacking**).
Unconfined aquifer	An **aquifer** where the water table is exposed to the atmosphere through openings in the overlying materials.
Waste	Any substance or object that the holder discards, intends to discard or is required to discard.
Wastewater treatment works	Installation to treat and make less toxic domestic and/or industrial effluent.
Watercourse	A natural or artificial linear structure that transports water (river, canals, culverts etc).
Water table	The level of groundwater in soil and rock, below which the ground is saturated. The water table may change with the seasons and following rainfall.
Well	An excavation that is drilled, bored or dug to locate, monitor, dewater or recharge groundwater.
Wetland	An area where saturation or repeated inundation of water is the determining factor in the nature of the plants and animals living there.

Abbreviations

ACT	Australian Capital Territory
ADR	European Agreement concerning the International Carriage of Dangerous Goods by Road (in force since 1968, latest revision in 2005)
BAP	biodiversity action plan
BOD	biochemical oxygen demand
BS	British Standard
CCS	Considerate Constructors Scheme
CCTV	closed circuit television
CCW	Countryside Council for Wales
CEEQUAL	Civil Engineering Environmental Quality Assessment and Award Scheme
CEH	Centre for Ecology and Hydrology
CDM	Construction (Design and Management) Regulations 1994
CIRIA	Construction Industry Research and Information Association
COD	chemical oxygen demand
COPA 1974	Control of Pollution Act 1974 [Scotland]
COPR	Control of Pesticides Regulations 1986
COSHH	Control of Substances Hazardous to Health Regulations 1988
CPO	compulsory purchase order
cSAC	candidate special area of conservation
CTRL	Channel Tunnel Rail Link
DARD	Department of Agriculture and Rural Development [Northern Ireland]
DBFO	design, build, finance and operate
Defra	Department for Environment, Food and Rural Affairs
DETR	Department of the Environment, Transport and the Regions
DfT	Department for Transport
DMRB	*Design manual for roads and bridges* (Highways Agency, 1998)
DO	dissolved oxygen
DoE	Department of the Environment [NI for Northern Ireland]
DTI	Department of Trade and Industry
DTLR	Department of Transport, Local Government and the Regions
EA	Environment Agency [England and Wales]
EA 1995	Environment Act 1995
EAP	environmental action plan
ECI	early contractor involvement

EIA	environmental impact assessment
EHS	Environment and Heritage Service [Northern Ireland]
EMS	environment management system
ENDS	Environmental Data Services
EPA 1990	Environment Protection Act 1990
ES	environmental statement
FCB	Fisheries Conservancy Board [Northern Ireland]
FEH	*Flood estimation handbook* (IH, 1999)
FEPA 1985	Food and Environment Protection Act 1985
FRS	Fisheries Research Services [within the Scottish Executive]
GBRs	General Binding Rules
HA	Highways Agency
HDD	horizontal directional drilling
HSE	Health and Safety Executive
IBC	intermediary bulk container
ICE	Institute of Civil Engineers
IDB	internal drainage board
IECA	International Erosion Control Association
IH	Institute of Hydrology [now Centre for Ecology and Hydrology]
ISO	International Standards Organization
KPI	key performance indicator
LPA	local planning authority
MAGIC	Multi-Agency Geographic Information for the Countryside
MCEU	Marine Consents and Environment Unit
MHWS	mean high water springs
MSDS	manufacturer's safety data sheet
NNR	national nature reserve
NERC	National Environmental Research Council
NTU	nephelometric turbidity units
OGC	Office of Government Commerce
ODPM	Office of the Deputy Prime Minister
PFI	Private Finance Initiative
PICP	pollution incident control plan
PPG	Pollution Prevention Guideline
PPE	personal protective equipment
PPP	public private partnership
RUSLE	revised universal soil loss equation
SAC	special area of conservation

SAM	scheduled ancient monument
SEA	strategic environmental assessment
SEPA	Scottish Environment Protection Agency
SMEs	small and medium-size enterprises
SNH	Scottish Natural Heritage
SPA	special protection area
SPZ	source protection zone
SSSI	site of special scientific interest
SUDS	sustainable drainage system(s)
TCPA	Town and Country Planning Act 1990 (as amended)
ToR	terms of reference
UKAS	UK Accreditation Service
WEWS	Water Environment and Water Services (Scotland) Act 2003
WFD	Water Framework Directive (2000/60/EC)
WRA	Water Resources Act 1991 [England and Wales]
WWTW	wastewater treatment works
μm	micron (ie 1×10^{-6} m)

Part A

Characteristics of linear projects and understanding water pollution

1	Introduction	29
	1.1 Surface water	29
2	**Water environments**	**31**
	2.1 Surface water	32
	2.2 Groundwater	33
	2.3 Water on site	36
3	**Water pollution and the law**	**37**
	3.1 Types and sources of pollution	37
	3.2 Pollution offences	38
	3.3 Water Framework Directive	40
	3.4 Responsibility for and costs of pollution	41

1 Introduction

There are more water pollution incidents from construction sites than from any other industrial sector in the UK, with 180 incidents from construction and demolition sites recorded for England and Wales in 2004 (<www.environment-agency.gov.uk>). At every stage of the construction process there is potential for water pollution problems to arise. Linear construction sites potentially pose a greater risk to the water environment because of the variety of environments they may affect, their cumulative impacts and the distances that require management.

1.1 CHARACTERISTICS OF LINEAR PROJECTS

In this guidance, the following types of linear projects are considered:

- roads – including new motorways, dual carriageways, bypasses, road widening, tunnels, bridges
- railways – including all new railway lines and on-line infrstructure track upgrades, tunnels, bridges, light rail systems, tramways
- pipelines – including new and replacement water, sewerage, oil, gas and chemical pipelines
- cables – including high- and low-voltage electricity supplies below ground or overhead, telecommunications cables
- watercourses – including canals, flood defences, river diversions.

By their nature, linear projects – roads, railways, pipelines, cables and watercourses – are usually large-scale schemes, often predominantly rural in nature. Linear construction projects differ from other construction projects in that they have dynamic site boundaries and cover large areas exhibiting varied physical characteristics. The number of watercourse crossings, discharge points, site compounds and haul roads are inevitably greater than on a static site. The route may cross varied environments, topography, soil types, geology and habitats etc, each requiring differing water management techniques.

Characteristics that distinguish linear projects from other construction sites include:

- a dynamic "corridor" of activity
- varying environmental and aquatic protection requirements in different areas
- numerous access points and haul routes
- cumulative impacts likely on a single watercourse or catchment
- greater variety of ground conditions and soil types
- restricted land take
- jurisdiction of differing regulatory authorities and trans-boundary issues
- longevity of schemes.

2 Water environments

Understanding surface water and groundwater environments is critical to:

- route selection
- scheme design
- planning construction working methods
- identifying mitigation measures to minimise the risk of water pollution.

Surface water and groundwater form two essential components of the water environment which, together with precipitation and evaporation form the hydrologic cycle (see Figure 2.1). Groundwater includes all water stored in permeable underground strata (or aquifers) and surface water includes watercourses, water bodies and runoff. The majority of these water bodies are legally termed "controlled waters" in England, Scotland and Wales, and "waterways" in Northern Ireland. In addition, some surface watercourses are designated as "main rivers", because of their importance in flood risk management. These are shown on statutory main river maps produced by the environmental regulators and government. It should be noted that the environmental regulators are responsible for all surface watercourses, including main rivers and smaller features.

A best practice holistic approach should be taken to consider and manage surface water and groundwater regimes as an integrated system – where surface water provides important recharge to groundwater and groundwater provides essential baseflow to rivers and wetland areas.

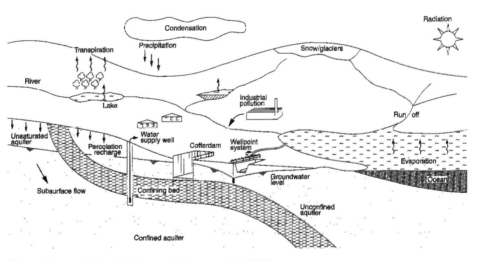

Figure 2.1　*The hydrologic cycle (Preene et al, 2000)*

Surface water and groundwater bodies are both highly vulnerable to pollution and impact by construction activities, especially in linear projects where there is high potential for the scheme to have multiple and cumulative impacts on the same water body or regime. Activities on construction sites located away from significant surface water bodies can still result in serious water pollution incidents, both by affecting adjacent small streams, ditches and drains that discharge to larger surface water bodies and by directly affecting the underlying groundwater regime. The effects of water pollution are costly and severe, often resulting in damage to water users and the natural environment some distance away from the polluting activity. Construction activities must not adversely affect surface water or groundwater features. Pollution incidents are often readily visible, are usually traceable to their source and are therefore likely to result in prosecution.

2.1 SURFACE WATER

The surface water environment includes:

- watercourses – natural and artificial, open and covered, including rivers, streams, storm drains, ephemeral ditches and canals
- water bodies – natural and artificial, for example wetlands, lakes and reservoirs
- runoff – controlled and uncontrolled.

Surface watercourses and bodies provide important water resources (for potable and other supply), general amenity and aesthetic value, recreational facilities (eg boating and fishing), conservation and ecological environments (see Chapter 24). Surface water can also provide recharge to groundwater systems.

All surface water bodies are vulnerable to pollution. Surface water pollution is usually readily visible and may reduce water quality, significantly change flow characteristics (level and volume), or significantly modify or destroy physical habitats (see Table 2.1).

Surface water pollution may result in prosecution and is likely to result in ecological damage, fish kill and loss of water supply, amenity value, and recreational use.

Table 2.1 *Construction activities that pose a high risk of surface water impact*

Pollution risk	Hazards
1 Activities that provide a pollution source	❖ Uncontrolled sediment erosion and contaminated silty runoff ❖ refuelling facilities, chemical and waste storage or handling areas ❖ polluted drainage and discharges from site ❖ contaminated groundwater from dewatering of contaminated sites
2 Activities that cause significant variations in natural flow	❖ Unregulated and poorly considered abstractions and discharges eg dewatering ❖ changes to the existing drainage network including interception and redirection of natural and artificial watercourses (eg field drains) ❖ discharge of groundwater to surface water ❖ increased runoff from cleared and capped areas (relative to greenfield values)
3 Activities that significantly modify or destroy physical habitats	❖ Watercourse crossings ❖ works within water ❖ outfall points

Large-scale linear construction projects are likely to encounter numerous natural and artificial water features and transect several watersheds or catchments. The water features encountered may be variable in:

- size, physical character and use
- flow/discharge characteristics and groundwater baseflow component
- conservation and ecological sensitivity
- catchment size, physical character and land use.

It is important to identify and characterise **all** surface watercourses and bodies at and near the corridor of the route, including rivers and streams, ephemeral ditches and field drains, foul and surface water drains and outfalls, canals and leats, lakes, ponds, reservoirs and wetlands and areas prone to flooding. It is also important to be aware that surface water flow (level and volume) is likely to vary with season and may show a flashy response to heavy rainfall. Surface watercourses that are dry or have very low flows during dry conditions may flow or even flood during winter and spring months or following heavy rainfall. In addition, surface water flows may vary as a result of adjacent abstractions and discharges. Understanding the potential variation in natural flows is important because:

- dry weather low flows will determine acceptable discharge quality standards and abstraction volumes – as natural dilution potential and sustainable abstraction quantities are lowest during periods of dry weather flows
- wet weather peak flows will determine the sizing of temporary culverts and structures, and acceptable discharge volumes – as the natural capacity for additional volume loading is lowest during peak flow events.

Key guidance

Further information on river flows (and groundwater levels) throughout the UK can be found in the National Water Archive, which is maintained by Centre for Ecology and Hydrology (CEH) at Wallingford (<www.nerc-wallingford.ac.uk/ih/nrfa/index.htm>).

Understanding the surface water environment is critical to minimising the risk of pollution by identifying vulnerable surface water features, likely impacts and developing sound mitigation measures. For example:

- physical characteristics of each watercourse will reflect the ground conditions and determine the natural form of the drainage channel and the need for appropriate mitigation measures (eg bank erosion control)
- watercourse depth and flow will affect the available dilution of discharges
- catchment size and potential runoff volumes will influence the size and location of balancing ponds (see Chapter 18)
- flooding potential will influence the location of site compounds and the programming and phasing of construction activities see Chapter 18).

2.2 GROUNDWATER

Groundwater occurs in permeable strata within the sub-surface. Groundwater is contained and flows within the pore spaces and fissure spaces of soils and rocks. Below the water table, the pore and fissure spaces are saturated; above the water table, within the unsaturated zone, the pore and fissure spaces contain a mixture of air and water. The permeability or hydraulic conductivity of a sub-surface unit is dependent on the size and connectivity of the pore spaces in granular units (eg gravels and sands) and the fissures in fractured units (eg granites and limestones). Units with high natural permeabilities are termed aquifers (eg chalk, sands and gravels) – classified as major and minor aquifers depending on their hydraulic properties, water resource potential and usage. In contrast, units with very low natural permeabilities that act as barriers to groundwater flow or are only capable of transmitting small amounts include very fine-grained units (eg clays and mudstones).

Groundwater is generally in constant motion and naturally flows along a hydraulic gradient from high points (where surface water, runoff and precipitation infiltrates the ground) to low points (where groundwater typically discharges and provides important baseflow to rivers and wetland areas). In addition, a proportion of groundwater is held in aquifer units deep within the sub-surface that have no natural discharge points. In an unconfined or water table aquifer the upper surface of the aquifer is exposed to the air and infiltration of water pollution. In a confined aquifer the upper surface of the saturated zone is sealed by a low-permeability unit that holds the groundwater at increased pressures within the aquifer. Therefore extreme care must be taken when drilling or piling through confining layers because rising, artesian, groundwater levels will rapidly be encountered, and a new pathway to the aquifer will be generated.

Groundwater is an important resource, providing more than one-third of the potable water supply in the British Isles. In addition, it provides essential baseflow to rivers and wetland areas, often supporting important ecological systems. However, groundwater is vulnerable to pollution – especially because it is generally less apparent than surface water and the potential impacts on groundwater are rarely observed and so tend to receive little consideration. Groundwater pollution is problematic because aquifer pollution persists for long periods and is often very difficult and costly to remediate: groundwater pollution prevention measures cost 10–20 times less than groundwater clean-up and aquifer remediation programmes. Groundwater quality is endangered by construction activities that provide a pollution source or pathway or that significantly vary natural groundwater levels (see Table 2.2). In contrast to surface water, groundwater is generally more vulnerable to pollution by chemicals, metals, hydrocarbons and salts than by sediments, because particulate pollutants are naturally filtered during infiltration and recharge. Pollution of groundwater is likely to result in the loss of potable or other water supplies, the degradation of receiving river or wetland waters and habitats, and, for offenders, prosecution.

Dewatering activities present a significant risk to groundwater through the following mechanisms:

- excessive dewatering in coastal and estuarine areas may induce saline intrusions
- dewatering may draw on to the site contaminated groundwater from off site and thereby generate a contaminated discharge
- dewatering may compromise the yield of nearby water abstractions.

In addition, artificial aquifer augmentation programmes may cause natural groundwater levels to rise, in turn causing remobilisation of contaminants within the unsaturated zone.

Table 2.2 *Construction activities that pose a high risk of groundwater impact*

Pollution risk	Hazards
1 Activities that provide a pollution source	❖ Fuel and chemical use and storage ❖ waste handling, storage and disposal ❖ accidental spillages ❖ use of concrete, bentonite and grout ❖ uncontrolled discharges ❖ works in contaminated land
2 Activities that provide a pollution pathway	❖ Tunnelling ❖ piling ❖ boreholes ❖ excavations
3 Activities that cause significant variations in groundwater levels	❖ Dewatering activities during excavations, earthworks, and tunnelling ❖ artificial recharge activities

Groundwater is likely to be present at depth along all routes, but the degree of vulnerability of different groundwater regimes to pollution may vary widely along a route and will be dependent on aquifer characteristics and the depth of the water table.

In general, groundwater is highly vulnerable to pollution in areas where:

- classified aquifer (major and minor) units are at outcrop with little (< 2 m) protective (silty/clayey) overburden deposits
- the water table is shallow
- there are known direct recharge pathways (eg limestone karst swallow holes).

Groundwater is also at a high risk of pollution in areas where it is directly encountered – ie when working at or below the water table in deep excavations, earthworks, tunnelling and piling. In these situations, a direct pathway to the aquifer exists with little or no natural protection.

Groundwater levels are likely to vary with season and may show a flashy response to heavy rainfall. Excavations that are dry in summer may require dewatering during winter and spring or following heavy rainfall.

The scheme may encounter numerous groundwater abstractions from boreholes, wells and springs. These sources may be operated by water companies for public supply or by private abstractors for domestic, industrial, agricultural or other use. Groundwater sources are most vulnerable to impact by construction activities within the catchment or recharge zone of the supply. The environmental regulators have defined source protection zones (SPZs) to protect groundwater sources that are used for public drinking supply. The SPZ framework is a risk-based classification that delineates sensitive areas where potentially polluting activities and/or accidental pollutant releases will have a detrimental effect on the quality or yield of a groundwater source. In general, the risk increases with increasing proximity of the polluting activity to the groundwater source. Construction activities must not affect the reliable yield or quality of any groundwater abstraction or receiving environment, and the acceptability of high-risk construction activities may be restricted within SPZs and high-sensitivity areas.

> **Key guidance**
>
> Boland et al (1999). *Guidelines and protocols for investigations to assess site specific groundwater vulnerability*
>
> EA (2005). *Groundwater protection: policy and practice*
>
> EHS (2001). *Policy and practice for the protection of groundwater in Northern Ireland*
>
> SEPA (2003). *Groundwater protection policy for Scotland*
>
> Geological and hydrogeological maps are available from the British Geological Survey, <www.bgs.ac.uk>.
>
> Information on groundwater levels (and river flows) throughout the UK can be found in the National Water Archive, maintained by Centre for Ecology and Hydrology (CEH) at Wallingford, <www.nerc-wallingford.ac.uk/ih/nrfa/index.htm>.

2.3 WATER ON SITE

All construction sites generate water from a number of sources. Any water that is wholly or partially produced in the course of construction activity or at a construction site is classed as "trade effluent". This includes water generated in the following situations on site:

- surface water runoff
- dewatering from excavations
- washing operations
- rainwater collected in drip trays and bunds
- road sweepers
- works in cofferdams
- wastewater from welfare facilities.

The only wastewater not regarded as trade effluent is:

- clean surface runoff (ie from roofs or hardstand car parks)
- clean water from dewatering operations (ie pumped groundwater).

Table 2.3 identifies some definitions of "water".

Table 2.3 *Definitions of water and wastewater*

Controlled waters (waterway in Northern Ireland)	All bodies of water, including groundwater, rivers, streams, canals, burns, lakes, lochs, coastal waters and tidal waters.
Trade effluent	Includes any water ("effluent") generated at construction sites, including surface runoff from buildings and hardstanding (as it is unlikely to be "clean"). The only effluents which are not classed as trade effluent are clean, uncontaminated surface water (ie clean rainwater that has not been contaminated when running over a site) and domestic sewage.
Sewage effluent	Domestic sewage/foul water.

3 Water pollution and the law

The law that governs the pollution of water is complex and continually evolving. Despite this, everyone in construction needs to be aware of the responsibilities imposed on them and of the consequences of any offence that might arise from their activities.

This chapter is not intended to be an exhaustive guide to the law, but it explains the key requirements in terms of water pollution. The responsibility for preventing pollution on site is discussed in Section 3.4 below. Clear and concise guidance on legislation relevant to several industries, including construction, can be found on the Netregs website, <www.netregs.co.uk>, run by the environmental regulators.

In addition to the various legal controls on water pollution itself, there are other related legal requirements:

- planning and development (Chapter 7)
- abstraction and other consents for construction (Chapter 13)
- wildlife protection (Chapter 24)
- handling and disposing of waste (Chapter 16)
- flood defence (Chapter 18)
- fisheries and works in water (Chapter 20).

3.1 TYPES AND SOURCES OF POLLUTION

Pollution has a number of legal definitions that now, importantly, include things which might lead to effects on ecosystems or people:

- poisonous, polluting or solid matter
- substances that harm the health of human beings or other living organisms,
- substances that harm the quality of the water environment, including aquatic and terrestrial ecosystems dependent on the water environment
- substances that cause offence to the senses of human beings
- substances that cause damage to property, and/or
- substances that cause impairment of, or interference with, amenities or other legitimate uses of the water environment.

Table 3.1 illustrates types of pollution and typical sources at construction projects.

Table 3.1 *Pollution types and sources*

Type of surface or groundwater pollution	Source at construction projects
Suspended solids – silt, sediments, "muddy" water	Surface water runoff, dewatering, outfalls, concrete, bentonite and grout operations, works in water
Organic compounds including hydrocarbons – fuel, oil, some chemicals	Contaminated land, fuel storage and use, vehicle maintenance, plant and vehicles, waste management
Metals – dissolved and in suspension	Contaminated land, dewatering
Salts and nutrients	Application and runoff of road salt, surface water runoff, application of fertilisers
Alkaline pH	Concrete, bentonite and grout operations
Microbial	Domestic/site sewage
Other chemicals	Herbicide and pesticide application, contaminated land, use of chemical additives (concrete operations etc)
Solid waste matter (litter, stone, wood, vegetation etc, particularly if it has the potential to block the flow of the river)	Uncontrolled waste handling and disposal, works in or near water, temporary watercourse crossings (haul roads)
Discoloration – as a result of clay, mud, chalk particles etc	Surface water runoff, dewatering, discharge, concrete, bentonite and grout operations, works in water
Energy – temperature	Well dewatering, recharge, overpumping, cooling, steam curing, washing operations

3.2 POLLUTION OFFENCES

Almost any uncontrolled discharge to a water body (including groundwater) has the potential to result in a criminal offence. Water pollution is an offence of strict liability – in all cases it does NOT require proof of negligence or that actual harm was done.

It remains an offence in England, Wales, Northern Ireland and Scotland to cause pollution to enter a controlled water, unless it is within the conditions of a discharge consent. Controlled waters (or waterway in Northern Ireland) include most inland waters, tidal water, groundwater and also ditches or streams that for the time being may be dry.

	England and Wales	**Scotland**	**Northern Ireland**
Offence	"cause or knowingly permit" pollution or solid matter to enter a controlled water without consent	"the direct or indirect introduction, as a result of human activity, of substances or heat into the water environment which may give rise to any harm" without consent	"knowingly or otherwise" discharge or deposit material in a waterway or groundwater without consent
Law	Water Resources Act 1991	Control of Pollution Act 1974	Water (Northern Ireland) Order 1999

While directly causing pollution is obviously an offence (eg pumping silty water into a river), permitting pollution through negligence is regarded with equal or greater concern – **ie knowing a source of pollution exists and doing nothing about it**. This is of particular importance on a linear site where site security is harder to ensure or where works may not be taking place on sections of the route during certain phases of the project. Potential pollution risks (such as off-site agricultural runoff or on-site runoff into site drainage) may still exist in these areas, which are the contractor's responsibility. Reporting pollution to the environmental regulator will be taken as mitigation.

> The law relating to water pollution may appear complex but the principle is simple: it is an offence to cause pollution.

Water entering road drains and surface water drains will discharge directly into controlled waters, possibly some distance from the site itself. On a linear site, the outfalls from surface water drains can be within a very wide area and should be located before work starts.

> **Construction firm fined £6000 for silt discharge** (source: *The ENDS Report*, Nov 2004)
>
> A civil engineering firm has been fined £6000 after admitting polluting a Hertfordshire brook with chalky silt. Several members of the public complained that the stream was coloured white. Environment Agency officers found that half a kilometre of the watercourse was affected and traced the pollution back to a construction site. Officers found that there was a deep shaft filled with cloudy, white water and staff admitted that they had been pumping water via a small settlement tank into a surface water drain. The drain led directly to the brook. The company were fined £6000 and ordered to pay costs of £1488.

An offence will have been committed if discharge takes place to a foul sewer without written permission from the local statutory sewage undertaker.

Further controls on groundwater pollution are imposed by the Groundwater Regulations 1998 (in England, Wales and Scotland) and the Groundwater Regulations (Northern Ireland) 1998, which control the discharge of certain substances listed in the EC Groundwater Directive (80/68/EEC). The Regulations prevent the discharge of List I substances (for example, oils, other hydrocarbons, pesticides, herbicides, cadmium, mercury) and control the discharge of List II substances (including certain metal compounds, aluminium sulphate, chlorine, zinc) to groundwater by requiring prior authorisation from the environmental regulator.

> **Key guidance**
>
> UK management guidelines on pollution of surface waters and groundwater: <www.netregs.co.uk>
>
> Groundwater Directive pages at <www.environment-agency.gov.uk>.
>
> Legal requirements, policies and guidelines on groundwater in Scotland: <www.sepa.org.uk/groundwater/lpg.htm>.
>
> DoE (NI) (1999). *Groundwater Regulations (Northern Ireland) 1998 Guidance Note 2*, <www.ehsni.gov.uk>.

3.3 WATER FRAMEWORK DIRECTIVE

The legal framework surrounding the water environment is undergoing the biggest change for years as a result of implementation of the Water Framework Directive (WFD). Over time, much of the existing European legislation will be repealed. New and amended legislation will expand the scope of water protection with the intention of achieving a "good" status (defined in terms of ecological, chemical and quantitative status in the legislation) for all waters by 2015.

The implementation of the WFD superimposes a river basin management planning framework and common environmental objectives for water bodies, on to the UK's existing system of water pollution controls (discharge consenting, abstraction licensing). Crucial to an understanding of the WFD is an appreciation of the fact that although it is concerned with "water" its scope is not restricted to what happens on or in water.

> The quality of any river body of groundwater or other body of water will be determined not just by what happens within its banks but also by what happens on the land around it. A spill of polluting material, eg oil, quite remote from a river, may well find its way there with devastating effects. The construction of a housing development may create serious flooding problems downstream.

For that reason, the scope of the WFD is not restricted to rivers, lakes or coastal waters – rather it requires consideration of any human intervention that could affect the quality of water, wherever that intervention takes place.

At the time of writing, the key pieces of enabling legislation of the WFD are as follows:

- **The Water Environment (Water Framework Directive) (England and Wales) Regulations 2003**
    - ❖ Water Act 2003 (in England and Wales)

    The Water Act establishes new regulatory arrangements for water abstraction and impoundment. It will be phased in by 2012; however, the majority of licensing changes came into force in April 2005. Three forms of abstraction licence are established – **transfer licences** (eg for dewatering engineering works, tunnels, road construction etc), **temporary licences** (less than 28 days) and full licences. Abstractions taking less than 20 m³/day do not require a licence. For further details on licences refer to Chapter 13.

- **Water Environment and Water Services (Scotland) Act 2003**
    - ❖ Water Environment (Controlled Activities) (Scotland) Regulations 2005

    The Controlled Activities Regulations came into effect on 1 October 2005 to control abstractions, discharges and certain engineering works for the purposes of protecting the water environment. The underlying principle is that no activity which affects the water environment should take place without prior authorisation. In summary, this framework is based on three tiers of control: **registration**, registration under **general binding rules** and **licences**. For further details on licences refer to Chapter 13.

- **The Water Environment (Water Framework Directive) Regulations (Northern Ireland) 2003**

    These address the preparation and implementation of river basin management plans. The regime for licensing and pollution offences under the Water (Northern Ireland) Order 1999 remains.

> **Key guidance**
>
> Water Framework Directive guidance is available from:
>
> <www.sepa.org.uk/wfd> for Scotland
> <www.defra.gov.uk/environment/water/wfd/> for England and Wales
> <www.ehsni.gov.uk/environment/waterManage/wfd/wfd.shtml> for Northern Ireland.

3.4 RESPONSIBILITY FOR AND COSTS OF POLLUTION

The contractor is responsible for management of water on the site. If watercourses become polluted or unacceptable materials are disposed of in sewers, the company responsible and/or the directors or any employee could be prosecuted. Where subcontractors or joint ventures are in place, their responsibilities should be defined by their contract (see Chapter 10), and may include any water management issues associated with their work activities.

Where a watercourse already appears polluted before construction, report it to the environmental regulator and obtain good baseline data (see Chapter 14).

It is illegal to breach the conditions of a discharge consent. Contractors should be aware that the named consent-holder is responsible for compliance with a consent. Should a subcontractor or other third party cause a breach of a consent, it is the consent-holder who will be prosecuted.

In all cases the polluter pays. Penalties for offences are maximum fine of £20 000 or imprisonment for up to three months, or both on conviction in a magistrate's court. In a crown court, the penalty is an unlimited fine or a prison term of up to two years, or both. In addition to this, the polluter will be required to pay the costs of clean-up, which can amount to far more than the fine. Often overlooked are the other hidden costs – for staff time, compensation, legal expenses, hire of external experts and delays to the construction programme.

The environmental regulator can serve a works notice requiring measures be taken to prevent a potential water pollution incident occurring. Failure to comply with a works notice is an offence and if the environmental regulator is required to undertake the works itself, the contractor will also be liable for the costs. Requirements for remediation of environmental damage are not restricted to prosecutions. The environmental regulators impose such requirements up to 10 times more frequently than prosecutions (Postle and Vernon, 2002).

While the contractor may be the one who is prosecuted for a pollution incident, the effects of delays, bad publicity and loss of reputation are likely to be felt by the whole project team and will reflect on the project promoter. It is therefore in the best interests of all to ensure that every care is taken throughout the project to prevent such an occurrence.

Pollution offences

Developer fined for polluting the River Lee – inadequate silt measures (source: EA press release, Apr 2005)

A contractor was fined £15 000 and had to pay the Environment Agency's costs of £1664.38 for allowing silt to enter a ditch tributary of the River Lee. An environment officer visited the site after a member of the public made a complaint about the quality of the water in the ditch.

The officer found that although filtration units were in place these were inadequate, allowing silt-laden water from the site to be discharged into the ditch. The company was informed that they needed a discharge consent to pump water into the ditch and that only clean, unpolluted water could be pumped into a river. It was agreed that additional "silt traps" in the form of hay bales were to be put in place.

Again the public informed the Environment Agency that brown water was entering the River Lee. An environment officer traced the flow upstream to the same site where he saw silt-laden water being discharged from a pump into the ditch. Although actual harm had not been assessed, the potential for harm could not be discounted.

Blocked sewer causes flooding (source: *The ENDS Report*, Nov 1995)

A contractor was fined £10 000 for two pollution incidents during construction of a bypass. Public complaints were received when sewage flowed over fields and into a brook. The company could not explain how the sewer became blocked with stones but admitted responsibility.

Blackburn Bypass pollution incidents (source: *The ENDS Report*, Nov 1995)

A contractor was fined a total of £7000 with £3000 costs for two pollution incidents during construction of the M65 Blackburn Bypass. The first occurred after a heavy vehicle fractured a water main, flooding the site with thousands of gallons of water. The company pumped the silt-contaminated water into the river without consent. The second offence occurred when a diverted stream was polluted after a pump malfunctioned.

Channel Tunnel Rail Link contractors were jointly fined £45 000 for polluting three Kent streams with silt (source: *The ENDS Report*, Feb 2001)

The first incident occurred when a member of the public reported a stream at Sandling running white with sediment. An Environment Agency officer found that the pollution originated from a CTRL construction site. Bales of hay and hessian had been placed in the stream but were ineffective in trapping the pollution.

The second incident, in February 2000, was near Maidstone. The Boxley Drain was heavily discoloured with silt originating from a CTRL construction site. Grossly contaminated rainwater was being allowed to run into the stream.

The third and most serious incident affected a designated wildlife site noted for its birds, grass snakes and amphibians. Members of an angling club told the Agency that the Musketstone Stream near Hollingford was so full of sand and silt that it was difficult to see where the original bank began.

The Agency told the court: "The defendant companies failed to put in systems properly designed to protect the watercourses from being polluted unnecessarily. Bearing in mind the size of the companies and the contract, more priority should have been given to anti-pollution measures."

The companies were jointly fined £15 000 for each of the three incidents and ordered to pay costs of £7000.

Part B

Planning and design

4	Introduction	45
5	Scheme design and land take	47
	5.1 Route selection	47
	5.2 Design	48
	5.3 Land take	49
6	Stakeholder consultation	51
7	Development consent	55
	7.1 Introduction	55
	7.2 Enabling and non-planning consultation	55
	7.3 Permitted development	56
	7.4 Planning permission	57
	7.5 Environmental impact assessment	58
8	Site investigations and monitoring	59
	8.1 Introduction	59
	8.2 Baseline monitoring	59
	8.3 Site investigation data used to manage pollution risk	60
	8.4 Pollution caused by site investigations	61
9	Programming and seasonal influences	63
10	Contracts	65
	10.1 Type of contract	65
	10.2 Tender and contract specification	66
	10.3 Liability	67

4 Introduction

Reducing the risk of water pollution from a construction site should start at the planning and design phase of a project, well before the construction starts. The planning and design phase is one of the most crucial of all the phases of a development; as it is during this phase that the fundamental decisions are made by clients, designers, contractors and regulatory authorities, which ultimately determine whether, and in what form, the project can proceed.

"Planning" in this context means the development of a scheme from its inception to readiness for construction, rather than the process of obtaining "planning consent", but can include that process (see Chapter 7).

> It is at the planning and design stage of a project that the potential for causing water pollution during construction should first be considered and either designed out or environmental management plans put in place and adopted throughout the construction phase.

During the planning and design phase the potential for causing water pollution during construction should be fully assessed. Key issues that must be considered are:

- route selection – avoiding constraints by route alignment (eg proximity to a water course)
- design (eg bridge design setting back works from the edge of a river)
- adequate land take (eg allowing sufficient land for temporary treatment works during construction)
- consultation with stakeholders (eg agreeing working practices with the environmental regulator and securing discharge consents)
- development consents (eg planning permission conditions)
- site investigation and monitoring (eg obtaining site-specific data to determine sensitivity and establishing baseline conditions as a benchmark before works begin)
- programming (at feasibility stage consideration of timing of works noting seasonal constraints, eg wet winter and spawning seasons for fish)
- contractual requirements (eg ISO 14001 and site environmental plans, specific contract requirements allowing for realistic pricing for pollution control measures).

The following chapters in Part B explore these issues in detail as they relate to linear construction projects and the control of water pollution during construction.

Planning any scheme should be an iterative process, whereby a client's initial proposals are presented, discussed with regulators and other interested parties, developed further, agreed, and then taken through to construction. The stages at which stakeholder consultation (eg environmental regulator) or a contractor becomes involved may vary. Figure 4.1 illustrates the stages in the development of a project where consultation with regulatory bodies and the involvement of contractors can help to minimise the risks of incidents occurring during the construction phase. The chart has been reproduced from River crossings and migratory fish: design guidance (Scottish Executive, 2000). The chart refers to a road or river crossing, but the procedures are equally applicable to route planning and detailed design of other works.

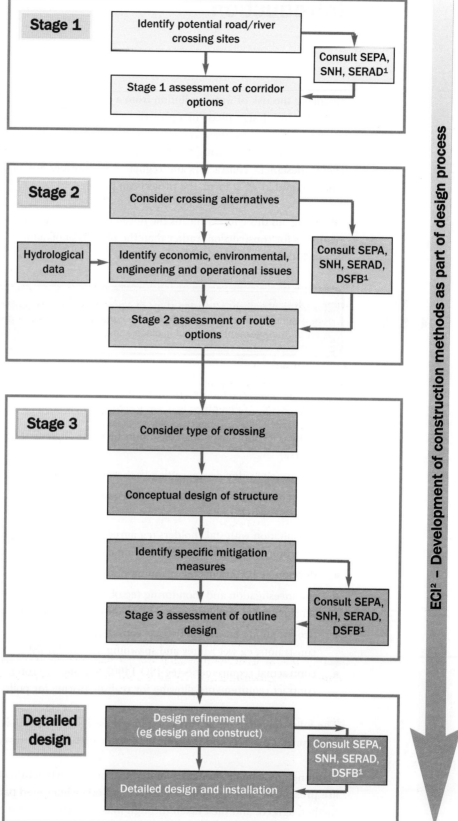

Figure 4.1 *Stages in the development of project proposals (after Scottish Executive, 2000)*

Note 1 – Consult environmental regulators, government organisations and conservation bodies as part of the iterative design process. Consultation to develop construction methods, agree timing of works and need for any licences.

Note 2 – Involvement of contractor as part of the design team helps to ensure construction solutions are practical, cost effective and address environmental issues (eg timing of works, method of construction and need for temporary works)

5 Scheme design and land take

5.1 ROUTE SELECTION

The selection of a route for a linear construction project is never as simple as drawing a line from the point of origin to the point of destination. Of primary importance are engineering feasibility, construction cost and availability of land. Environmental factors also need to be taken into account when selecting a route.

Many of the environmental constraints to planning a route not only affect engineering design and scheme cost, but also have a bearing on construction methodology and the potential to create, or avoid causing, water pollution. In terms of the water environment, areas of particular sensitivity and land type should be identified as early as possible in the route selection process and avoided where possible. See Chapter 2 for further information on types and characteristics of water environments. Key considerations include:

- major aquifers
- source abstraction zones
- watercourses of high chemical or biological quality
- ecologically sensitive wetlands or watercourses
- contaminated land
- topography and soil type (runoff potential).

Collection of all the constraints data would normally be undertaken by the design consultants in the preparation of the design. It is essential that these data are passed on to the contractor and are used to prepare environmental management plans for the construction phase.

The Multi-Agency Geographic Information for the Countryside (MAGIC) website allows easy identification of designated or otherwise sensitive areas and land classifications, such as SSSIs (sites of special scientific interest) and nitrate-vulnerable zones. However, it is recommended that information on statutory designations for use in route planning, decision-making and site management be collected from the relevant authority (eg environmental regulator and conservation body websites).

> **Key guidance**
>
> **Selected sources of information on the water environment**
>
> Environment Agency's indicative flooding maps in England and Wales (at present Scotland and Northern Ireland do not have such maps).
>
> MAGIC website, <www.magic.gov.uk>.
>
> Lists of local, national and European designated sites such as sites of special scientific interest (SSSIs) held by the conservation bodies (Natural England, CCW, SNH).
>
> The Department for Transport's web-based Transport Analysis Guidance, webTAG, for road schemes in England and Wales, <www.webtag.org.uk>.
>
> Consultation with stakeholders (see Chapter 6)

As well as aiming for the most appropriate horizontal alignment to avoid sensitive sites, the vertical alignment also needs to be considered. Projects involving cuttings and embankments can pose an increased risk of water pollution by creating stockpiles and expanses of sloping surfaces exposed to surface water runoff (see Chapter 18).

Where significant environmental constraints exist, the best practice would be to select a route that avoids them altogether. Where it is not possible to avoid them, measures to prevent, reduce or even compensate for any adverse effects will normally be required. These are commonly called environmental mitigation measures.

5.2 DESIGN

The main objective is to ensure that the fundamental engineering decisions avoid creating water pollution hazards, particularly when surface water or groundwater conditions have been identified as being sensitive to pollution. Table 5.1 provides some examples of design controls. These are simplistic examples, suggesting alternative approaches to design decisions. They may not be suitable in all cases, and the decision needs to balance environmental protection with engineering buildability and cost.

Table 5.1 *Examples of design controls to avoid water pollution hazards*

Source of hazard	Examples	Control options
Existing contaminants	Earthworks and/or foundation design	Leave undisturbed, remove under controlled conditions – perhaps ahead of main work; treat the land to reduce exposure.
	Foundations	Use driven piles (or even raft foundations) in place of bored piles
Specified materials and work processes	Solvent-based coatings	Specify water-based products or other alternatives
	Concrete structures such as outfalls or bridge piers	Prefabricate off site; cast on site in designated area away from water
Specified design	Bridge over watercourse	Extend bridge to locate piers "inland" rather than in or adjacent to watercourse.
	Works adjacent to or in water	Design works to be accessed and constructed from one bank
	Bank reinforcement	Use stone, gabions, wooden planks or geotextiles rather than concrete or sheet piles
Construction method	Cable or pipeline route	Route underground rather than above ground (depending on groundwater sensitivity)
	Cutting/embankments	Design for shallow slopes where possible and in all cases design adequate pre-construction cut-off and slope drainage
	Bridge crossing	Design to work within watertight cofferdams rather than in open water, or bunded areas

Design can also be used to build-in features that enhance the water environment, for example by including sustainable drainage systems (SUDS) to clean up runoff in a more "natural" way than a traditional engineered solution.

It can be seen from these examples that such decisions may impose additional construction costs. These should, however, be compared to the additional costs of mitigation measures otherwise required on site as well as the increased potential for prosecutions. Such decisions may also help to enhance a scheme's chance of approval by the regulators, users and the public.

5.2.1 Design standards

Many client organisations have design standards that incorporate guidance related to water pollution. While these are often aimed at the operational phase of a scheme, some include construction aspects. Some examples are:

- Highways Agency (1998), *Design manual for roads and bridges* (DMRB)
- Network Rail (nd), *Design manual for rail drainage systems*
- London Underground Limited (nd), *Guidance on drainage design, installation, maintenance and removals*
- British Waterways (2005), *Code of practice for works affecting British Waterways*
- environmental policy commitments, implemented via a range of procedures or codes of practice, eg the design and construction of new water or utility facilities.

5.3 LAND TAKE

Linear schemes are typically constructed within a restricted corridor of available land. A common challenge faced by such projects is the requirement for additional land to manage, contain and treat water on site. Many of the measures recommended in Chapters 18 and 19 to control water from linear construction sites require land, eg for constructing a settlement pond.

> One of the key requirements for successful pollution prevention is to ensure that there is sufficient room on-site to contain and treat water before discharging it from site.

Land take can be either permanent (eg the land taken for a new road) or temporary. Temporary land take is generally land required during construction, which is then restored to the original or an agreed alternative use (eg temporary land take to install a pipeline or for construction compounds for any scheme).

The extent of permanent land take and construction corridors will be known and the effects of the works assessed as part of the development consent process (see Chapter 7). When planning and designing the route, consider the requirements for additional temporary land take for water pollution control including settlement ponds, diversion ditches, buffer strips, temporary oil separators as well as site compounds and storage areas. Where constraints plans have been developed as part of route planning and design, consult these to identify the optimum location for any additional temporary works. Involve contractors in assessing construction methods at an early stage to allow a more realistic picture to be developed of the overall land take requirements.

In determining land take requirements for water pollution prevention, it is necessary to evaluate which areas of the project will generate water and require storage or water treatment during construction (see Chapters 18 and 19 for more information). In some cases, this will be necessary for the permanent works, such as balancing ponds for highway drainage. Among the situations where this may apply are:

- cuttings and embankments requiring top drainage
- tunnels and shafts
- dewatering
- excavations in former or existing mining areas that may cause mine-water break-out
- construction of bridge piers or other structures within cofferdams in open water
- low-lying areas to where runoff will gravitate
- stockpiles
- other activities that may dam or restrict surface waters.

Early identification of these additional sites will allow time for consultations with landowners and agreements to be reached before construction begins. If an environment impact assessment (EIA) is required, the location of settlement ponds, site compound(s) and other temporary land take should be assessed as part of this process. While there are clear benefits to using additional land to create water pollution measures during the construction phase, a balance must be struck as these works can have other impacts (eg including disruption to existing land-use, cost of land purchase or rental, impact on any buried archaeology and impact on nature conservation and landscape features).

Where possible, designers and contractors should plan a phased construction programme that allows the temporary works (eg ponds and buffer strips) to be located within the permanent land take area. Also the design and construction team should seek to optimise the use of permanent works (eg permanent ponds and or ditches) as part of the temporary construction control measures. Both approaches will limit the amount of additional land required.

If the client already owns the land that is to be taken, it is considerably easier to proceed with the project. By obviating the need to acquire ownerships (through either voluntary or compulsory purchase), the time and expense of legal inquiries, financial negotiations and eventual land and property purchase are avoided. Where work is taking place on land that the client already owns, as in the case of improvements within existing road or rail boundaries, space should be allowed for water storage and treatment.

The Highways Act 1980, Transport and Works Act 1992 and Harbours Act 1964 grant clients powers of compulsory purchase to allow them to acquire land. Local planning authorities can, given appropriate justification, compulsorily acquire land for a project. These powers are derived from the Planning and Compulsory Purchase Act 2004 and are termed compulsory purchase orders (CPOs).

Where there is insufficient land available to accommodate construction-related activities, and where compulsory purchase is deemed to be an unnecessarily expensive or drastic measure, an alternative is to rent land from landowners near to the proposed project alignment.

> In all cases, planning or other forms of consent applications (including environmental impact assessments) (see Chapter 7), will all need to address temporary land take and construction activities on and off site in the same way that that they will need to address the construction and operation of the project itself.

6 Stakeholder consultation

Consultation with stakeholders is an essential part of the development of any scheme. In addition to the project team, stakeholders include:

- statutory consultees, such as the Environment Agency, SEPA and Natural England. It is a legal requirement to consult them on planning or other development applications. Most of the statutory consultation requirements are set out in Article 10 of the Town and Country Planning (General Development Procedure) Order 1995

- non-statutory consultees, comprising public and private bodies such as the Wildlife Trusts, as well as certain functions of otherwise statutory bodies, which should be consulted. Many of the non-statutory recommendations for consultation are set out in Appendix B to Circular 9/95 (DoE, 1995), but other circulars and planning policy guidance notes refer to further bodies

- other consultees, including landowners, businesses, water users downstream and the public, who can be consulted where appropriate.

Table 6.1 lists the more common stakeholders and their statutory or non statutory interest in the water environment. The regional or local office of these organisations would normally handle the consultation response. It should be recognised that the regulators have statutory obligations to perform and will be able to advise on any requirements, standards or guidelines with which the project must comply. Contact details of the regulators and other organisations can be found under "More resources" at <www.netregs.gov.uk>.

Contact the relevant regulators and statutory bodies etc as early as possible in the scheme development.

It is particularly important to consult stakeholders where the design or environmental assessment of the project needs to take account of sensitive environmental issues, eg the presence of a major aquifer or a designated site such as an SSSI or SPA.

Local or specialist ecological groups may hold valuable information or monitoring data, and should therefore be consulted. This can save time and costs that may otherwise be incurred carrying out necessary monitoring or employing a specialist consultant.

Consultations should cover a wider area than just the route corridor. Where a route crosses a river, the whole catchment should be considered, and downstream users – eg water abstractors or fisheries – some distance from the route should be identified and consulted. Similarly, underground aquifers can be extensive and may support numerous private or commercial water supplies, all of which need to be identified. See Chapter 2 for more information on the water environment.

Table 6.1 *Consultees with interests in protecting the water environment*

Organisation	Statutory responsibility (to water environment)	Interest in water environment
Environment Agency (England and Wales), <www.environment-agency.gov.uk> The Environment Agency has area and regional offices throughout England and Wales. To contact a local office, telephone 08708 506506. Emergency hotline: 0800 807060	Water resources/abstractions Surface water quality and pollution Groundwater quality and pollution Contaminated land Licences for works near water, discharges to surface water or groundwater, abstractions	Navigation Conservation Flood risk and defence
Scottish Environment Protection Agency (SEPA), <www.sepa.org.uk> SEPA has 21 offices across Scotland and a corporate office in Stirling, tel: 01786 457700 Emergency hotline: 0800 807060.	Water resources/abstractions Surface water quality and pollution Groundwater quality and pollution Licences for works near water, discharges to surface water or groundwater, abstractions	Conservation Flood risk and defence
Environment and Heritage Service Northern Ireland (EHS), <www.ehsni.gov.uk> Water Quality Unit, tel: 028 9025 4754 Emergency hotline: 0800 807060	Surface water quality and pollution Groundwater quality and pollution Waste management Drinking water supplies Licences for discharges to surface water or groundwater, abstractions	Conservation and designated sites
Rivers Agency (Northern Ireland), <www.riversagencyni.gov.uk>	Drainage Flood risk and defence	Navigation Conservation
British Waterways, <www.britishwaterways.org.uk>	Navigation Landowner	Amenity Water quality Abstraction Ecology
Internal drainage boards (IDBs) (England and Wales). Association of Drainage Authorities <www.ada.org.uk>	Land drainage Flood risk	Conservation
Navigation authorities – various including Environment Agency, local authorities, canal trusts and port authorities. Association of Inland Navigation Authorities, <www.aina.org.uk>	River amenity and navigation (where not under control of British Waterways)	Conservation Amenity
Local water company/sewage undertaker Regional water companies operate in England and Wales; in Scotland there is one water authority, Scottish Water; Water Service serves the whole of Northern Ireland.	Water supply Sewerage Licensing and policing disposal of trade effluent to sewers	Conservation
Natural England (formerly English Nature). Aat time of writing: <www.defra.gov.uk/ruraldelivery>. **Scottish Natural Heritage**, <www.snh.org.uk> **Countryside Council for Wales**, <www.ccw.gov.uk>	Conservation, designated sites, landscape Licences for works in/near designated sites; protected species licensing	Conservation
Private water abstractors		Water supply
Private/commercial fisheries		Watercourses
Landowners/riparian owners		Various

Public consultation often gives an opportunity to provide information about the project and to allay any concerns. This can be carried out by means of public meetings, exhibitions, leafleting, letter drops, advertisements and articles in the press, circulating questionnaires (where opinions are being sought) and posting information on a project website (the latter can be set up to allow people to post their responses online). Where possible, a central point of contact should be provided. For preference this should be someone who has sufficient project knowledge to be able to respond to queries immediately or to obtain information quickly from the project team. This may be the client's customer service team, a project PR manager, the project manager or site manager for example.

Clients and contractors sometimes regard a consultation exercise simply as an invitation (for local interest groups in particular) to raise objections to the scheme. Care should be taken to ensure that consultees understand why they are being consulted and that their views will be considered as part of a detailed decision-making process. Those consulted should have a clear idea of the kind of comments they are being asked to provide and within what timeframe. Unnecessary or unclear consultation wastes time and can cause delay or disruption. Early consultation ensures that parties are involved throughout the scheme and it can help significantly in gaining support and statutory approval, as well as reducing pollution incidents and subsequent prosecutions.

> **Key guidance**
>
> **Stakeholders**
>
> Stakeholder guidance is available from:
> - planning pages of <www.odpm.gov.uk>
> - <www.planningportal.gov.uk> for England and Wales
> - <www.scotland.gov.uk/topics/planning_building> for Scotland
> - <www.planningni.gov.uk> for Northern Ireland

7 Development consent

7.1 INTRODUCTION

Most linear construction projects require planning permission or some other form of development consent from a responsible authority before construction can start. The process and legislation governing the development process is not covered in this guidance. The main planning and enabling vehicles are discussed under the following headings:

- enabling and non-planning legislation
- permitted development
- planning permission
- environmental impact assessment.

Planning or other development consent applications are made to the responsible authority (or authorities). The responsible authority depends on the nature of the proposed development and the legislation under which the consent application is made.

7.2 ENABLING AND NON-PLANNING LEGISLATION

Enabling legislation governs the consent and development process for the particular type of project proposed. Linear projects largely represent public utilities or transport infrastructure and so generally have a specific public benefit. For these reasons, many are carried out under their own enabling legislation, such as the Highways Act 1980, Roads (Scotland) Act 1984, Electricity Act 1989 or Transport and Works Act 1992. Some of these acts also enable the approved development to proceed in other ways, such as by granting the client powers of compulsory purchase.

Such legislation may specify measures that must be carried out to control, among other issues, water pollution. For example, Schedule 9 of the Electricity Act 1989 (as amended) requires electricity companies to have regard to protection of amenity, flora and fauna in formulating any proposals. Generally, each act is accompanied by a circular or guidance note that sets out how to comply with the terms of the act, including relevant consent and environmental impact assessment requirements (see Key Guidance).

7.2.1 Transport and Works Act orders

Since 1993, new railway, tram or inland waterway projects can be authorised by orders made under the Transport and Works Act 1992. A Transport and Works Act order grants planning consent without the need for a formal application under the Town and Country Planning Act 1990. This exemption does not prevent action being taken against the works under other legislation such as that relating to statutory nuisance. The orders will often contain clauses for the protection of the water environment, for example the Transport and Works (Model Clauses for Railways and Tramways) Order 1992 provides a model clause on the discharge of water. For many schemes clients should still seek planning permission to ensure that environmental issues are considered and approved.

7.2.2 Major projects

Major new projects such as the Channel Tunnel Rail Link are generally authorised by private acts of Parliament. Such acts "disapply" or "substitute" the normal planning requirements in respect of the project concerned but do not usually override any requirements of environmental legislation. The Channel Tunnel Rail Link Act 1996, Schedule 2, includes requirements to take reasonably practicable steps to ensure that any water discharged is as free as may be practicable from gravel, soil or other solid substance, or oil or matter in suspension.

7.3 PERMITTED DEVELOPMENT

Railway, road and utility companies have permitted development rights that allow them to undertake some construction work without having to apply for planning permission. The conditions and limitations for this are set out in the Town and Country Planning (General Permitted Development) Orders applicable to the relevant region of the UK.

The right to permitted development is restricted by the regulations for environmental impact assessments (EIAs). If a development requires an EIA, then the permitted development rights are overruled and planning permission must be sought.

If environmental regulators or conservation bodies are not consulted, projects deemed "permitted development" risk causing unintentional water pollution or environmental harm. However, the requirement for contractors to obtain the necessary licences for working in or near water, for dewatering, abstracting or for discharging water **still apply** to permitted developments (see Chapter 13 for details on licences and consents).

> "Permitted development" projects still require licences for construction activities including works in or near water, dewatering, water discharge or abstraction.
>
> Permitted development rights may be withdrawn if the scheme exceeds certain size criteria or if it is located in an environmentally sensitive site or requires an EIA.

Many utility companies avoid the risk of delay or prosecution by engaging in proactive informal involvement with environmental stakeholders in the development of new schemes (see also Chapter 6).

> **Case study – water supply pipeline** (source: Hyder)
>
> A proposed water supply scheme comprised a water abstraction, a 20 km supply pipeline, associated pumping stations and a new electricity supply. The scheme was progressing as "permitted development". Part of the pipeline crossed rural land that was a designated European protected site, so the client, designer and contractor met with the planning authorities, environmental regulators and conservation bodies early in the scheme to discuss suitable working methodologies.
>
> In reviewing the scheme as a whole it was found that both the pipeline and the water abstraction would potentially affect the European site. Consequently, permitted development rights were withdrawn and the whole scheme had to undergo an EIA and planning permission was required. This ensured the impacts of the scheme as a whole were addressed and subsequently mitigated, and the project team avoided prosecution.

7.4 PLANNING PERMISSION

For most developments for which the consent processes in Sections 7.2 and 7.3 do not apply, construction can only start once planning permission under the Town and Country Planning Act 1990 (as amended) has been obtained. National and regional planning policy is set out by the government in circulars, planning policy guidance notes and regional policy guidance. Protection of the water environment is considered in several planning policy guidance notes.

> **Key guidance**
>
> **Planning policy guidance notes and circulars**
>
> PPGs and Circulars are available from:
> - planning pages of <www.odpm.gov.uk>
> - <www.planningportal.gov.uk> for England and Wales
> - <www.scotland.gov.uk/topics/planning_building> for Scotland
> - <www.planningni.gov.uk> for Northern Ireland

Planning applications and supporting information are submitted to the respective planning authority. Where linear projects cross administrative boundaries, more than one planning authority may be involved in the decision-making process. When this occurs, the planning authorities concerned should reach an agreement that one authority takes responsibility for the planning application process. Normally this will be the one in which the majority of the proposed scheme is located. However, each authority must determine its own planning application for that part of the scheme which falls within its administrative boundary.

The application should adequately describe not just the nature of the scheme but also any construction activities and temporary land take requirements for site compounds, material and stockpile storage and settlement ponds etc (see Section 5.3 for guidance on land take).

> **Make sure all temporary land take for settlement ponds, access and drainage etc is included in the initial planning application, as it may be difficult to get further permission for additional land once consent has been granted.**

Individual planning applications are judged against the local development plan but the planning authority must also consider any other material considerations. These can include representations from the public and from statutory consultees (which often include the environmental regulators). To avoid delays, ensure that the issues which are likely to be of concern to the consultees are addressed and discussed with them from outset of project development. See Chapter 6 for further information on stakeholder consultation.

After an eight-week period of consideration (16 weeks for a proposal that requires an EIA), the local planning authority either refuses the application or issues a planning permission with certain conditions attached to it. These conditions must be complied with either before construction can begin or during construction, depending on the nature of the condition. Conditions to protect the water environment may include a programme of water quality monitoring, the submission of a site drainage plan before work starts or adherence to certain guidelines or methodologies such as the environmental regulator's pollution prevention guidelines.

7.5 ENVIRONMENTAL IMPACT ASSESSMENT

Environmental Impact Assessment Regulations require the environmental effects of certain projects to be assessed prior to permission for development. Legislation on EIAs for various types of development is extensive (see Appendix 1), but the "EIA Regulations" refer to the following:

- The Environmental Impact Assessment (Scotland) Regulations 1999
- The Town and Country Planning (EIA) (England and Wales) Regulations 1999
- The Planning (Environmental Impact Assessment) Regulations (Northern Ireland) 1999.

Developments are classed according to the impact that they are likely to have. For Schedule 1 projects an EIA is mandatory; Schedule 2 projects only require an EIA if they are liable to have a significant impact on the environment because of their size and/or location. Among other considerations, 2 km is the threshold for linear developments such as overhead cables, waterways, roads and railways (5 km for underground pipelines).

If a development requires an EIA then an environmental statement (ES) must be submitted in conjunction with the planning application. The requirements for this are set out in Schedule 4 of the EIA Regulations, which, importantly, requires an assessment of the impacts of construction and construction land take on the water environment.

Typically, the EIA will consider the existing (baseline) conditions, the potential impacts of the proposed project and mitigation measures that can be taken to minimise these impacts. Such mitigation measures must be adopted throughout the development of the scheme. Mitigation may apply to the scheme design phase: the route itself, constructing tunnels rather than bridges, using prefabricated structures rather than cast in situ; or to the construction phase: monitoring water quality, provision of water treatment facilities or the methodology for watercourse crossings.

> The environmental statement often sets out measures to mitigate water pollution during construction. They should be identified, planned for and followed throughout the scheme design and construction phases.

7.5.1 Appropriate assessment

If the project is likely to affect a European-designated site such as a special area of conservation (SAC) or a special protection area (SPA), the "competent authority" must complete an appropriate assessment. Depending on the scheme, the "competent authority" may be the nature conservation bodies or the environmental regulator, for example. The information required for an appropriate assessment is often provided in the environmental statement.

8 Site investigations and monitoring

8.1 INTRODUCTION

Site investigations are undertaken for all developments to identify ground conditions, influence route selection and establish design criteria. Site investigation information, obtained for design purposes, can also be used in the control of pollution.

Data obtained before construction begins provides a baseline against which the effects of construction can be compared. The investigation and monitoring work should also be used as the basis for comprehensive risk assessment and planning mitigation measures to control water pollution during construction.

In very sensitive environments, it may be necessary to undertake site investigations specifically targeted to the avoidance of pollution. In such circumstances, as with all site investigations, care should be taken to prevent pollution from the investigation works themselves.

8.2 BASELINE MONITORING

> It is essential to know the status of surface water and groundwater before construction starts. Mitigation measures should be designed to protect these baseline conditions in the water environment. Baseline data can then be used as a benchmark to determine what effect, if any, construction activities are causing.

A baseline survey of surface water features should include the presence, water quality, depth and flow characteristics of all water bodies at or near the site. Particular attention should be given to identifying ephemeral ditches and field drains that tend only to flow in wetter conditions and may be easily overlooked during site survey work. Groundwater data is usually obtained from the engineering ground investigation or a dedicated groundwater monitoring investigation, and can include water quality and water level as well as ground permeability and/or porosity.

The level of information obtained should be risk-based, depending on the likely impacts to occur and the sensitivity of the water feature. In some circumstances it would be adequate to obtain samples and test for a limited range of parameters such as suspended solids, hydrocarbons, BOD and pH. Where there are greater risks, for example the presence of contaminated land, then testing for a full suite of chemical determinands may be more appropriate.

Similarly, the timeframe over which testing is carried out must be related to the length of the proposed project and the anticipated risks. Water quality and, in particular, water levels will vary seasonally. For these reasons, it is not unusual for baseline ground and surface water monitoring to be conducted over one or more years. Further information on monitoring during construction is provided in Chapter 14.

> **Case study – monitoring for pipeline on SSSI** (source: Hyder)
>
> The fields and light industrial areas of the Gwent Levels SSSI are criss-crossed by a number of ancient drainage ditches whose waters support rare flora and fauna. A new sewerage pipeline crossed part of the SSSI. The Countryside Council for Wales guidelines for development within the SSSI required water sampling to be carried out quarterly before, during and for a year after construction of the scheme. Each sample was obtained following an agreed protocol and tested for a suite of some 20 chemical determinands based on the pollution that may occur from construction activities and disturbance of the ground, including pH, chloride, nitrates (fertilisers), BOD, dissolved oxygen, hydrocarbons and metals. Chemical concentrations detected in each sample were compared to agreed standards; in certain locations the baseline samples provided a useful and more realistic benchmark for assessing the water quality. The monitoring programme highlighted seasonal variations in water quality and short-term effects of some watercourse crossings, but demonstrated that no long-term impact had been caused by the pipeline construction.

Where a programme of monitoring is not implemented, publicly available data accumulated over as long a period as possible should be reviewed (such as river level data). In the case of flooding, records going back many years may be necessary to establish trends in flood levels and flood "return periods" (see Chapters 18 and 20). For estuarine or tidal river developments, tide tables are essential, and tidal records over time may show changing trends.

8.3 SITE INVESTIGATION DATA USED TO MANAGE POLLUTION RISK

The following data obtained from site investigations can be invaluable in the control of water pollution.

Water table depth and ground permeability

Groundwater characteristics (Chapter 2.2) will inevitably influence the design of major excavation, tunnelling and dewatering schemes. The rate of inflow of water into excavations, the volume of water requiring disposal and the potential for recharge can all be assessed. See Chapter 21 for guidance on dewatering.

Where monitoring of the water table is required over a period of time, the boreholes and equipment installed during the site investigation (eg standpipes and piezometers) are often maintained during and following construction. Where this is not the case, it is important to be aware that the timeframe of a site investigation typically allows only a snapshot of water level variability.

> Because groundwater levels usually vary according to rainfall or time of year, the potential exists for higher (and lower) water levels and greater water volumes than may be calculated from data obtained during the brief site investigation. Monitor water levels over as long a period as possible.

Depth to groundwater and the permeability of the ground are key elements that determine the suitability of land for infiltration, either to assess the potential for runoff (Chapter 18) or to dispose of pumped silty water (Chapter 19).

Existing groundwater quality

Having baseline chemical analyses of the existing status of the groundwater is essential to determine whether construction is having any impact on the water quality. Monitoring regimes before, during and after construction are often required for major projects that could affect groundwater (see Chapter 14).

Water quality data can also influence the design of dewatering schemes, particularly the suitability of the water to be discharged to a watercourse, sewer or other option (see Chapter 21). Chemical data may be required when submitting an application for permission to discharge the water (see Chapter 13).

Presence and characteristics of ground or groundwater contamination

Where land contamination has been identified, earthworks and temporary surface water drainage measures must be contained within the contaminated area and not mixed with soil or water from the rest of the site. If contamination has been identified, record the boundaries and depths in any environmental plans, surface water management plans (and waste management plans) for the site and design activities such as earthworks, dewatering and control of surface water runoff particularly to deal with the contamination. Guidance on dealing with contaminated land is provided in Chapter 23.

Susceptibility of ground to erosion and sediment generation

Soil and rock types directly affect the suspended sediment (silt) pollution that may occur in excavations or surface water runoff. Fine-grained soils or rock such as chalk, clay and shales are easily eroded and sediment particles require a long retention time to settle out, thus requiring a large storage area or alternative management strategies. Guidance on runoff can be found in Chapter 18 and on removing silt pollution in Chapter 19.

Ground conditions for trenchless construction

Directional drilling is becoming a popular method of crossing watercourses. The more comprehensive the results from the site investigation, the greater the chance of significantly reducing the risk of a bentonite breakout during drilling. Directional drilling contractors, who are experienced in identifying the best borehole locations, should be involved. Guidance on trenchless construction is provided in Section 20.4.

8.4 POLLUTION CAUSED BY SITE INVESTIGATIONS

Site investigations are typically excluded from routine construction operations, largely because they necessarily take place well in advance of any form of construction. Nevertheless, the risk of water pollution still exists. The following **risks** should be assessed by technical specialists before mobilising an investigation:

- drilling boreholes through the base of a lined landfill, creating a pathway for contamination into uncontaminated groundwater
- drilling boreholes into areas of land contamination, creating pollution pathways
- drilling boreholes into or through an aquifer, resulting in potential effects on water quality and resource
- site investigations in or over water
- site investigations on or near protected sites
- vehicular access to exploratory hole locations, including vegetation clearance, causing polluted runoff from exposed, muddy or rutted ground
- supply and use of water for rotary drilling or for purging boreholes
- use of bentonite and other drilling muds.

Key guidance

Geological and hydrogeological maps available from the British Geological Survey, <www.bgs.ac.uk>

BS 10175:2001 *Investigation of potentially contaminated sites. Code of practice*

BS 5930:1999 *Code of practice for site investigation*

Boland et al (1999), *Guidelines and protocols for investigations to assess site specific groundwater vulnerability*, <http://publications.environment-agency.gov.uk/pdf/SPRP2-042-01-e-p.pdf>

9 Programming and seasonal influences

Most development schemes are driven by target dates for such things as financial expenditure, planning permission, tender award and project completion. Within these timeframes adequate allowance must be made to plan for and implement the various requirements to control water pollution. The following information should be read in conjunction with that in Chapter 12, which relates to construction programming.

Planning and design

Formal consultation, planning permission or other consenting processes have statutory timescales, as discussed in Chapter 7. Where "informal" consultation with regulators takes place, for example for a permitted development, sufficient time should be allowed for them to respond and, where necessary, for their comments to be discussed and addressed in the design of the scheme. It is worth noting that they have six weeks to respond to formal consultation on a scheme under the various Town and Country Planning Regulations.

Monitoring and surveys

Adequate time should be allocated for the identification and monitoring of water features and groundwater before works begin. The effect of seasonal variations on scheme construction should be assessed and several months monitoring may be necessary. The benefits of baseline monitoring are discussed further in Section 8.2. If an adequate time allowance is not made at an early stage, the planning authority or the environmental regulator may require monitoring work to be completed before construction can start. If there has been no allowance for monitoring in the pre-construction programme, costly delays may ensue.

Ecological surveys, where required, can only be carried out at certain times of the year, depending on the habitat or species in question. Early identification of the appropriate time windows by a specialist ecologist can avoid delays later on in the scheme. Further information on ecology is provided in Chapter 24.

Licences and consents

Written permission is required from the environmental regulator or other authority to undertake certain construction activities such as discharging water to a watercourse or sewer, carrying out works in a watercourse, abstracting or dewatering. In many cases a licence can take up to four months to obtain, so if it is required at the start of construction, submit the necessary applications well in advance of construction. In some cases, clients or their representatives can apply for a licence and subsequently transfer it to their nominated contractor after award of tender. Details of licence requirements are provided in Chapter 13.

Temporary works design

Planning to minimise disturbance from seasonal and unpredictable weather patterns will enable successful programming of the construction works (site planning is addressed in Chapter 12). It is usual for major earthworks to be programmed for the summer. Where a project is scheduled to last one or more years, this is often not possible. Flooding, too, is often seasonally related and is an increasingly important issue for all forms of development. The design of all temporary water management controls, and the land required to implement them, must be adequate for a rainfall event of an appropriate magnitude to the length of the scheme and its location. See Chapters 18 and 19 for guidance on water management controls, rainfall and runoff.

Construction programming

Chapter 12 provides information on the following construction programming issues which, if considered early enough in the scheme development, could be avoided or taken into account during the design:

- seasonal and geographic constraints
- water treatment methods
- dewatering
- water discharge
- working in winter
- ecological constraints
- water amenity
- landscaping and reinstatement.

10 Contracts

The contract between a client and their chosen consultant and/or contractor is the single most important means of communicating specific water pollution prevention requirements and expected standards. Communication of this information at bidding or tender stage as part of the scheme brief or terms of reference (ToR) is essential, as it is at this point that sufficient costs can be set aside and programmes and/or method statements developed that will allow appropriate measures to be built into the project. Likewise, main contractors should include specific requirements in subcontract documentation wherever possible.

10.1 TYPE OF CONTRACT

Contracts between clients and contractors can either be specific to the client organisation or standard forms of contract can be adopted. The Law Society can provide advice and guidance on forms of contract.

CIRIA C532 *Control of water pollution from construction sites. Guidance for consultants and contractors* (Masters-Williams *et al*, 2001) reviews the different types of contract between the client and contractor (this still applies at the time of publication). The guidance relates to the following forms of contract: traditional, design and build, design build finance operate (DBFO) and partnering.

In recent years Private Finance Initiative (PFI) and public private partnership (PPP) contracts have been used for certain government projects, including roads. In July 2002, guidance entitled *Green public private partnerships* (ODPM *et al*, 2002) was published on how to incorporate environmental considerations into PFI or PPP projects.

Under traditional forms of construction procurement, the contractor is not involved at all in the design and development of the project: it simply constructs what an engineer employed by the client has designed. While design and build contracts have been used, the contractor is often not involved until after planning approval has been obtained. At this stage the design is all but finalised and there is little scope for innovation and consideration of temporary works (including pollution prevention) or buildability.

Early contractor involvement (ECI) enables benefits to be gained from the contractor's skills much earlier in the project cycle. ECI allows contractors to be involved in the completion of the statutory procedures as well as the subsequent design and construction of the works.

The benefit of having a contractor in place at this early stage is not only to take advantage of the contractor's expertise in construction methods but to provide more scope for innovation, improve risk management, better planning and programming, better quality and value for money. Furthermore, with the increasing environmental considerations it is imperative the contractor understands and is party to delivering the environmental and other commitments given during the scheme development process.

10.2 TENDER AND CONTRACT SPECIFICATION

When tendering for work set out in a project brief or terms of reference (ToR), which will form part of the contract on award, contractors have the opportunity to provide information as to how they would control pollution. While contractors have a legal obligation not to cause pollution, if the specific contract requirements are not set out at this point, it is left to each contractor to determine the importance of this aspect in their submission.

1 If the tender is to be determined on price alone and the ToR are non-specific about pollution prevention, the incentive for a contractor to price in measures or allow for contingency/risk items for pollution is limited.

2 If the tender is to be determined on quality, the contractor can make written statements of intent, but again if the ToR are not specific about the issue, the contractor's response may be written in a way that can later prove difficult to implement.

3 A tender may be determined on past performance data (PPD). In this case it would be for the contractor to demonstrate its environmental awareness, performance and commitment to the control of pollution. However, PPD on pollution prevention would normally only be provided if the ToR were specific on this issue.

Regardless of how the tender is determined, it is unlikely when considering all other aspects of a tender submission that a contractor will focus on pollution prevention unless the ToR mention it specifically.

> One of the main drivers for environmental improvements is pressure applied by clients through standards laid down in contract documentation.
>
> Contracts should specify exact requirements for water pollution prevention in order to encourage high standards and to allow like-for-like tender evaluation.

Although the statutory duty not to pollute falls on all developers and operators regardless of the development consent or any contract conditions, the client or client's adviser (traditionally the consultant) should determine the need for specific contract requirements in relation to pollution control.

Clearly the detail of any specific contract clause will depend on the particular circumstances of the project. If the work in the planning and design phase has demonstrated that the type of work has a risk of causing pollution and/or the water environment is sensitive it is not adequate merely to state "…the Contractor shall implement all necessary measures to prevent pollution…". In making such a broad requirement, it:

- limits the information provided to the contractor on the particular measures that may be required or expected at any one site
- limits the contractor's ability to innovate or propose beneficial solutions that may not be cost-competitive at tender stage, but which would prove invaluable later on
- may result in a lack of resources available on site for adequate pollution prevention measures
- may result in the preferred contractor (by price) not offering the best solution for pollution control.

> As a minimum, best practice requires that the contract documentation:
>
> - includes specific clauses relating to the control of water pollution
> - highlights potential risks relating to water pollution identified at the planning and design stage
> - requests project documentation such as environmental management plans, erosion plans, pollution contingency plans or codes of practice where risks have been identified
> - cross-references any specific requirements set out in the development consent (eg planning permission conditions) or the environmental statement (see Section 7.5)
> - provides all available information held by the client such as water quality monitoring and survey data
> - schedules any relevant codes of practice that should be complied with such as the environmental regulator's pollution prevention guidelines (PPGs)
> - specifies who is responsible for obtaining any necessary licences and consents (client or contractor) and includes details of any that have been obtained. The quality of any river body of groundwater or other body of water will be determined not just by what happens within its banks but also by what happens on the land around it. A spill of polluting material, eg oil, quite remote from a river, may well find its way there with devastating effects. The construction of a housing development may create serious flooding problems downstream.

In addition to setting out the rights and duties of the parties, the contract can be supported by specifications, briefs, drawings, reports (such as the ES), development consent conditions and codes of practice. There may be statements in the ES or planning consent etc that are effectively binding on the contract. This is because consent for the project will only have been granted on the basis of the application details submitted. The client must therefore ensure that commitments or conditions made are implemented by specifying them as mandatory in contract clauses.

Where such documents do not exist, specific clauses should be included relating to such matters as:

- providing detailed method statements for and subsequent approval of the design of temporary and permanent surface water drainage
- preparation of and approval of pollution contingency plans, sediment and erosion control plans and construction method statements for specific operations
- management of sewage effluent (from site compounds)
- monitoring of surface water quality at and adjacent to the route before, during and after construction
- monitoring of groundwater levels and quality before, during and after construction

10.2.1 Environmental plans

Many major construction companies and a growing number of SMEs have ISO 14001-certified environmental management systems, within which their sites are required to operate. Information on these systems, such as evidence of the environmental policy or certification, can be requested at tender stage. It is increasingly common for contractors to be asked to provide site environmental management plans, in draft at tender stage and then developed following award of contract and throughout construction. Further information can be found in CIRIA C533 *Environmental management in construction* (Uren and Griffiths, 2000).

Specific documentation can be required depending on the circumstances – many highways schemes require a erosion and sediment control plan; Network Rail often requires a pollution incident control plan (PICP) to be prepared in advance through its

Contract requirements – environment (Railtrack, 2002). The production of sediment and erosion plans has long been a requirement of major projects in the USA and Australia, from where much of the available published guidance originates.

10.2.2 Codes of practice and agreements

Some projects agree with the regulators a construction code of practice that tends to be project-specific (see case study below), or is specific to certain activities, such as a watercourse crossing, supplemented by best practice measures for fuel use etc.

A requirement to follow industry codes of practice – including for example the environmental regulators' Pollution Prevention Guidelines (PPGs) (see key guidance below) or CIRIA SP156 *Control of water pollution from construction – guide to good practice* (Murnane *et al*, 2002) – can be included as a contract requirement.

Where clients have specific requirements, these too should form part of the contract agreement. For example, British Waterways requires contractors to follow its *Code of practice for works affecting British Waterways* (BW, 2005), while Network Rail's project- and contractor-specific operating agreements set out guidelines for contractors on a range of activities including the use of herbicides, the response to pollution incidents and requirements for track and ballast cleaning.

Controlling spillages on the CTRL (source: Osborne and Montague, 2005)

An example of good practice includes the procedures developed for the construction of the Channel Tunnel Rail Link (CTRL). The project involved building more than 88 km of new track on highly vulnerable aquifers that are exploited for public water supply. An increased risk of spillage during the construction was identified. A risk assessment included an estimate of the total frequency of spillage of materials that could cause groundwater pollution. Accordingly, mitigation measures were developed, such as the provision of sealed drainage, to minimise the impact of any spillages occurring. The good practice procedures were included in all contract documentation.

The Civil Engineering Environmental Quality Assessment and Award Scheme (CEEQUAL) is an awards scheme assessing the environmental quality of civil engineering projects. Its objective is to encourage the attainment of environmental excellence in civil engineering projects, and thus to deliver improved environmental performance in project specification, design and construction. A CEEQUAL award for a civil engineering project publicly recognises the achievement of high environmental performance.

CEEQUAL uses a credit-based assessment framework and includes environmental aspects such as the use of water and land as well as ecology, nuisance to neighbours, waste and community amenity. Awards are made to projects in which the clients, designers and contractors go beyond the legal and environmental minima to achieve distinctive environmental standards of performance. Clients can therefore specify that a scheme apply for CEEQUAL.

Clients can also register for the Considerate Constructors Scheme and contractors can include it in their subcontracts. The scheme is a voluntary code of considerate practice that is adopted by any participating construction companies and everyone involved on the construction site. The code commits those contractors in the scheme to a range of standards including being considerate neighbours, clean, safe environmentally conscious and responsible. The code specifically requires that attention be paid to the avoidance of pollution. National awards are presented to the best-performing sites.

> **Key guidance**
>
> Pollution Prevention Guidelines available to download from <www.environment-agency.gov.uk/ppg>, <www.sepa.org.uk> and <www.ehsni.gov.uk>, for example:
>
> PPG1 *General guide to the prevention of water pollution*
>
> PPG5 *Works in, near or liable to affect watercourses*
>
> PPG6 *Working at construction and demolition sites*
>
> Uren and Griffiths (2000), C533 *Environmental management in construction*
>
> Masters-Williams *et al* (2001), C532 *Control of water from construction sites. Guidance for consultants and contractors*
>
> British Waterways (2005), *Code of practice for works affecting British Waterways*, <www.britishwaterways.co.uk/images/COP_2005.pdf>
>
> Details about CEEQUAL and registration, <www.ceequal.com>
>
> Details about CCS and registration, <www.considerateconstructorsscheme.org.uk>
>
> European and US guidance on sediment and erosion plans are listed in the References, pp 225–230

10.3 LIABILITY

Where subcontracts are in place on a project, it may be assumed that the main contractor will take responsibility for licences, water management and pollution control. Environmental law is not that simple, however. If a subcontractor causes pollution, intentionally or accidentally, it can be prosecuted even if the main contractor is contractually bound to manage any outfall and drainage etc. The main contractor may also be prosecuted for "knowingly" allowing pollution.

A similar situation may also arise where the main contractor is, for example, the named holder of a discharge consent and is thus responsible for ensuring the quality of the water discharged meets the conditions of the consent, irrespective of whether a subcontractor causes polluted water to be discharged.

It is sometimes the case that a client applies for a licence in advance of appointing a contractor to allow the necessary timeframe for a licence to be obtained. On tender award, the licence can then be transferred to the contractor (by informing the environmental regulator in writing) for the duration of the construction.

It is recommended that the environmental regulator is consulted and/or professional legal advice is sought for further advice and guidance.

> While financial liabilities and responsibilities can, and should, be specified in controls – ie stating who is responsible for drainage, water management, obtaining necessary licences – pollution remains a case of strict liability irrespective of any contract.

Part C

Construction

11	Introduction	75
12	**Site planning**	77
	12.1 Introduction	77
	12.2 Environmental management plans	77
	12.3 Risk assessment and control	78
	12.4 Consultation with regulators and other organisations	79
	12.5 Programming and seasonal pollution control issues	80
	12.6 Roles and responsibilities	83
	12.7 Emergency procedures	85
13	**Licences and consents**	87
	13.1 Discharging to sewer	87
	13.2 Discharging to surface water or groundwater	88
	13.3 Abstracting and dewatering	89
	13.4 Works in or near water	90
	13.5 Works in tidal waters	92
14	**Monitoring**	93
	14.1 Legal requirements	93
	14.2 Benefits of monitoring	93
	14.3 What to monitor	94
	14.4 When to monitor	94
	14.5 How to monitor	95
	14.6 Records	98
15	**Emergency and contingency planning**	101
	15.1 Risk assessment	101
	15.2 Emergency plans and procedures	102
	15.3 Training and testing	103
	15.4 Equipment	104
	15.5 Corrective action	105
16	**Site set-up**	107
	16.1 Introduction	107
	16.2 Site drainage and water features	108
	16.3 Water supply	110
	16.4 Water use	110

	16.5	Wastewater disposal 112
	16.6	Storage and use of materials 113
	16.7	Waste management 116
	16.8	Fuel and oil 117
	16.9	Site security 122
17	**Adjacent land and water use** **125**
	17.1	Protecting adjacent land and water uses 125
	17.2	Protecting the site from adjacent activities .127
	17.3	Additional land take 128
18	**Runoff and sediment control** **129**
	18.1	Introduction 129
	18.2	Preparing an erosion and sediment control plan 130
	18.3	Estimating runoff 132
	18.4	Flooding 137
	18.5	Estimating sediment generation 138
	18.6	Erosion and sediment control measures 140
	18.7	Protecting existing and pre-construction drainage 151
	18.8	Sustainable drainage systems (SUDS) 154
19	**Water treatment methods and disposal** **155**
	19.1	Introduction 155
	19.2	Sediment 155
	19.3	Concrete and cementitious material 167
	19.4	Fuel and oil 167
	19.5	Metals 169
	19.6	Ammonia and oxygen levels 170
	19.7	Sewage 170
	19.8	Disposal options and temporary outfalls 170
20	**Works in or near water** **175**
	20.1	Planning the works – legal requirements 175
	20.2	Pollution controls 176
	20.3	Access and haul routes across water 179
	20.4	Trenchless construction 181
	20.5	Open excavations and diversions 184
	20.6	Overpumping 186
	20.7	Bank works 187
	20.8	Works near watercourses 188
	20.9	Works in the floodplain 190
	20.10	Works over water 190
21	**Excavations and dewatering** **193**
	21.1	Legal requirements 193
	21.2	External dewatering (groundwater) 193
	21.3	Internal dewatering (excavations) 194

22	**Concrete and grouting activities**	199
	22.1 Legal requirements	199
	22.2 Alternative methods	199
	22.3 On-site batching	200
	22.4 Transport and placement	200
	22.5 Tunnelling, thrust-boring and pipejacking	202
23	**Contaminated land**	205
	23.1 Introduction	205
	23.2 Investigation and assessment	205
	23.3 Development of specific mitigation	205
	23.4 Managing unexpected contamination	206
24	**Ecology**	209
	24.1 Legal protection	209
	24.2 Construction impacts	211
	24.3 Vegetation clearance and landscaping	212

11 Introduction

Despite the controls put in place during the design and development of a project, there are many activities for which water pollution can only be prevented through appropriate construction methods, integral pollution controls, awareness and understanding of the issues and well-considered contingency measures.

This chapter identifies the potential causes of water pollution that may arise from construction processes and provides guidance on the management and control of water on site. Chapters 12–19 provide guidance on planning and setting up site, licence requirements, controlling runoff, water treatment and general pollution controls. The remaining chapters (Chapters 20–24) address some key construction activities and other important issues.

Figure 11.1 lists typical construction activities and the water pollution issues that may arise from them. Select an activity from those listed on the left-hand side of the table and read across to identify the chapters containing guidance. The guidance is graded in terms of relevance and importance.

KEY

▲ Important guidance
● Additional guidance

Note: Emergency and contingency planning (Chapter 15) is relevant to ALL activities on site

Activity	12 Site planning	13 Licences and consents	14 Monitoring	16 Site set-up	17 Adjacent land and water use	18 Runoff and sediment control	19 Water treatment and disposal	20 Works in or near water	21 Excavations and dewatering	22 Concrete and grouting activities	23 Contaminated land	24 Ecology
Abstraction	●	▲	▲	●	●		●		▲		●	
Bridges	▲	▲	●			▲		▲	●	●		●
Chemicals				▲								
Cofferdams	▲	▲	●				▲	▲	▲	●		●
Commissioning project		▲					▲					●
Compounds, site	▲			▲	●							
Concrete, cement and grout				●			●	●		▲		
Culverts		▲	●				●	▲	●	●		●
Decommissioning/demobilisation	●			●	▲							▲
Dewatering	●	▲	▲		●		▲		▲		●	
Directional drilling	●	▲					●	▲		▲	●	
Drainage works		●	●	▲	▲	●						
Dredging works	●	▲	●					▲			●	▲
Earthworks	●			●		▲			▲		▲	
Flood defences and bank works	●	▲	●			●		▲				▲
Floodplain, works in/on	▲					▲		▲				●
Fuel storage and refuelling				▲			●	●				
Haul route				▲	●	▲		●				●
Live systems	▲		●		●		●					
Material storage				▲	▲						●	
Monitoring/testing water	▲	●	▲				●		●			
Piling	●						●			▲	▲	
Pipejacking	●	▲					●	▲		▲	●	
Plant storage/maintenance				▲		●						
Reinstatement	●			●	▲							▲
Stockpiles				●	▲						●	
Topsoil stripping	▲					▲		▲				
Tunnelling	●	▲					●	▲		▲	●	
Vegetation stripping	▲					▲		▲				
Waste (solid and/or liquid)		●		▲			▲					
Watercourse crossing, works near or over	▲	▲	●			●		▲				●
Water discharge/disposal	●	▲	▲		●		▲	●				
Water treatment	●	▲		●	●	●	▲					

Figure 11.1 *Construction activities and water pollution issues discussed in this guidance*

Select an activity from the list on the left side of the table. Read across the table to identify the important issues (▲) and additional information (●) identified along the top of the table with the corresponding chapter number. For major construction activities, eg tunnelling, there will be a range of supporting activities, eg dewatering, which should also be cross-referenced.

12 Site planning

12.1 INTRODUCTION

The development and implementation of project management plans, site environmental procedures and work method statements is an integral part of the site planning process and provides controlled transition from the design phase of a scheme to construction. Good site planning is essential for the success and smooth running of a project. The planning process is a team effort and may require input from the client, designers, contractor, major suppliers and subcontractors on the scheme. The process can be significantly improved by working in partnership with other major stakeholders such as environmental regulators, statutory conservation bodies and water users. Planning on linear construction projects is much the same as on any other project. However, the route may cross several regulatory boundaries, different members of the project team will manage remote work areas along the length of the scheme, and certain areas of the route will only be active during certain phases, so a consistent approach is important.

12.2 ENVIRONMENTAL MANAGEMENT PLANS

Any environmental management system should use management resources efficiently and effectively to deliver the following objectives in a way that can be easily audited:

- compliance with environmental legislation likely to affect the project
- management of the major environmental impacts of the project
- provision of figures and data for environmental performance measurement
- systematic monitoring, review and improvement.

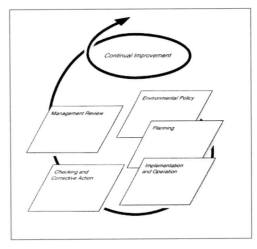

Figure 12.1 *ISO 14001 elements*

An environmental management plan is an effective way of employing the principles of environmental management systems (EMSs) at site level. Project environmental plans may be standalone or integrated with other management systems such as health, safety or quality to avoid duplication between the disciplines. Plans provide an overview of environmental management on site and will direct the team to more detailed information in site procedures or method statements. The level of detail the plan provides will depend upon the complexity and size of the project. A typical plan however, should describe arrangements for the following:

- aims, objectives and targets
- legal requirements and obligations
- organisation and responsibilities
- environmental risks

- management/control and mitigation
- training
- communication
- inspection and monitoring
- emergency arrangements
- review.

The international standard for environmental management systems is BS EN ISO 14001: 2004. Information on the benefits of EMSs and how to implement and audit one can be found in CIRIA C533 *Environmental management in construction* (Uren and Griffiths, 2000). Clear and concise guidance on legislation relevant to various industries, including construction, can be found on the Netregs website, <www.netregs.co.uk>, run by the environmental regulators.

12.3 RISK ASSESSMENT AND CONTROL

The risks to surface waters and groundwaters from construction works need to be identified at an early stage. On large schemes potential risks are often identified during the environmental impact assessment (EIA) (see Chapter 7) and reported in the environmental statement (ES) or environmental action plan (EAP). Environmental risks may also be identified during the design stage in the designer's risk register and decisions taken to revise the design, method of construction, or specified materials to reduce the risk. See Chapter 2 for information on water features that may be at risk.

On schemes where an environmental statement has not been prepared or risks have not been identified, a qualitative risk assessment or appraisal should be carried out.

When carrying out an initial risk assessment it is also useful to apportion the risk, for example to the client, contractor or landowner, and confirm who will be responsible for managing it during the various phases of the project. One key example is the risk of polluted off-site runoff entering, and polluting, site drainage.

Monitoring is an essential part of managing risk. Before works begin it is important to establish a baseline for the condition of water bodies around the site. These data can then be compared with the results of any monitoring carried out during construction and can be used to determine whether the work has had any adverse (or positive) effect on the water environment, particularly should there be a pollution incident. Regular inspection and monitoring of site activities are essential during the construction phase of the project. Further guidance on monitoring is provided in Chapter 14.

Early assessment of potential risks and the control measures to be used will enable the project team (including subcontractors) to incorporate the measures into the works and price for them accordingly. Further information on managing water pollution risk is provided throughout this guidance.

	Management best practice – risk assessment and control	
1	Identify the location of all sensitive receptors within or adjacent to the site	Mark up a site plan with the locations of all rivers, ponds, streams, aquifers, boreholes, field drains, fisheries, ecologically sensitive sites, surface and foul drainage systems. This will help with the overall planning of the site set-up (eg storage areas, refuelling points, haul routes, washout areas).
2	Identify sensitive receptors off site or downstream of the scheme (eg abstraction boreholes, fishery) that could be affected by the works	Undertake monitoring prior to construction to establish the baseline water quality and levels; make arrangements to continue this during and after the construction works (Chapter 14).
3	Identify construction activities and sources of pollution that may affect the water bodies identified	These could include piling, excavations, dewatering, river crossings, as well as general sources of pollution such as fuel storage, surface water runoff and concrete use.
4	Evaluate the risk of the construction activities polluting the water bodies identified	Assess the likelihood of an activity causing pollution. Assess the significance of the harm potential pollution would do to a particular water body (eg polluting an aquifer used for potable supply would cause greater potential harm than polluting a foul drain)
5	Implement mitigation to eliminate or reduce the risks	Use the following options to manage the risks: • remove the risk (eg use a different construction method; substitute certain substances or materials for less hazardous ones) • control the source (eg modify the construction method, provide adequate bunding of fuel and other storage areas, install measures such as silt fences or ditches to control runoff) • protect the receptor (provide hardstanding for compounds and storage areas where groundwater is sensitive, leave grass unstripped along a river bank to filter runoff, obtain necessary consents, licences or permissions from environmental regulators) • put procedures in place for emergency incidents

12.4 CONSULTATION WITH REGULATORS AND OTHER ORGANISATIONS

As early as possible contact should be made with the environmental regulator, other statutory bodies and stakeholders such as landowners, businesses and other water users downstream to discuss the project and advise them of the activities taking place. A list of typical stakeholders and their contact details can be found in Chapter 6.

Regulatory and statutory bodies welcome an early approach and will be able to advise on issues of local importance, constraints and the need for licences or consent. They may also advise on alternative methods of work, products and equipment, the need for pollution prevention measures and monitoring requirements. This is particularly relevant for projects where an EIA or prior consultation has not been undertaken.

It is important to note there are different departments within regional offices of the regulatory organisations that deal with discharge licences, pollution control, land drainage, abstractions and water resources. A linear project may cross one or more regional boundaries, and a single point of contact in any one department may not be appropriate. Where this is the case, it is important to contact each of the relevant offices and departments to ensure that all parties agree with the proposals (see Chapter 6).

12.5 PROGRAMMING AND SEASONAL POLLUTION CONTROL ISSUES

Many construction projects will be subject to specific timing restrictions such as ecological or seasonal constraints in addition to financial, land access and completion targets. These constraints often determine when construction activities may or may not be carried out and early identification of these is critical for developing the construction programme and sequencing the works (particularly when programming critical-path activities). Programming issues during the planning and design stage of a project are covered in Chapter 9. The following sections address issues that are specific to planning the construction phase.

> Careful programming and phasing of construction works can significantly reduce the risks of pollution to watercourses and delays in construction activities.

12.5.1 Seasonal and geographical constraints

Rainfall patterns can vary immensely geographically, seasonally and annually throughout the UK. Planning to minimise disturbance from often unpredictable weather patterns will contribute to successful programming of the construction works (see also Chapter 9). Long-range weather forecasts should be consulted and weather conditions monitored before works begin. Earthworks and similar construction activities should be programmed to avoid inclement weather and reduce the potential for pollution. Northern and western areas of the UK experience significantly higher rainfall than southern and eastern areas (see Figure 12.2). In the UK the winter is generally wetter than the summer, although there can be considerable annual variations, with unpredictable high and low rainfall years. High rainfall events are increasingly being experienced in the summer rather than the winter. In some regions this variation can result in summer droughts or in serious flooding.

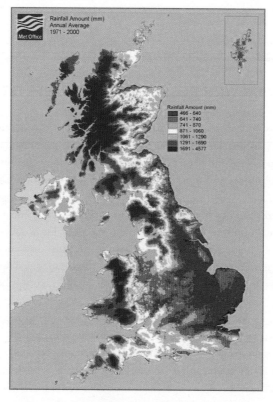

Figure 12.2 *Rainfall amount (mm) annual average 1971–2000 (Met Office, <www.met-office.gov.uk>)*

12.5.2 Water treatment methods

Temporary water treatment facilities established in the summer months may be inadequate to cope with greater volumes of water in the winter. Temporary works should be designed for the worst-case scenario at the outset (a 1 in 10-year storm is recommended). See Chapter 19 for details of water treatment methods.

Some forms of water treatment methods will also be affected by the timing of works. For example, the effective use of reed beds or swales as a final treatment system for highway drainage requires vegetation to be well established. When developing the construction programme it is important to consider the types of treatment methods and activities being utilised and to allow sufficient time for them to be established.

12.5.3 Dewatering

The timing of groundwater works is important, as groundwater levels will vary throughout the year depending on rainfall. A direct result of dewatering operations is the removal of groundwater from the surrounding area and the creation of new hydraulic gradients. Where abstraction (dewatering) is carried out during periods of sustained dry weather the activity may cause decreased groundwater levels, leading to a potential drought effect in surface watercourses as well as supply boreholes. Abstraction can also lower water levels some distance from the works.

Only the minimum volume and rate needed to produce suitable working conditions (dry excavations) should be dewatered. Where the risk of lowering surface water and groundwater levels around the site exists, water levels should be monitored during the works. A scheme of recharging water may help to mitigate any effects of dewatering; this should be discussed in advance with the environmental regulator and conservation bodies. See Chapter 21 for further guidance on dewatering.

12.5.4 Water discharge

During prolonged periods of dry weather, water pollution incidents are more likely to be severe. Low water levels in watercourses mean that there is less dilution of any substances discharged into the water. This has knock-on effects for the disposal of water from the site. A decrease in water levels is also usually associated with a decrease in water flow, exacerbating pollution impacts.

In dry weather it is particularly important to inspect the discharge outfall regularly to confirm that water treatment methods are effective and that the discharge is not causing scour or undercutting of the exposed bank profile. See Chapter 19 for further guidance on water treatment and disposal.

12.5.5 Working in winter

Winter is typically the wettest season. During this time, the ground can become so saturated that a high proportion, if not all, rainfall becomes runoff. Water control measures must be adequate to cope with this volume of water (see Chapter 18).

Road salt is often stored at sites with large compounds, car parks and access roads so that it is readily available for application in freezing conditions. While a valuable safety measure, road salt can cause serious water pollution if not controlled. Localised environmental damage from salt comes largely from stockpile runoff. There are also associated costs in lost material when materials are stored in the open. Stockpile runoff can be avoided by storing the salt under cover on an impermeable base. Proprietary storage bins are widely available.

Rock salt is specified in BS 3247:1991 *Specification for salt for spreading on highways for winter maintenance*. Concentrations of the chemical components of rock salt, sodium and chloride, can increase in surface and groundwaters. Other "chemical" de-icing agents are also used. Oxygen levels in water polluted by chemical de-icing agents are reduced as the materials biodegrade, which may endanger animals and insects living in the watercourse.

Salt is activated when crushed by vehicle tyres on a damp or wet road surface to form a brine solution, which lowers the temperature at which ice will form, but it is only effective if it can form a solution with the water on the road surface. If this water has already frozen before the salt is applied, the salt is less able to combat the slippery conditions (see salting rates in Table 12.1). Occasionally, sand is added to the salt, both to aid grip and to help break up existing ice. After the ice hazard has passed, this sand can cause problems with blocking drainage channels. Guidance is provided by the Highways Agency in its *Trunk road maintenance manual* (volume 2, part 3 "Winter maintenance code"), by the Scottish Executive's *Code of practice – winter maintenance for trunk roads* and by the Association of Metropolitan Authorities in its publication *Northern Ireland highway maintenance – a code of good practice, winter maintenance supplement*.

Table 12.1   *Rate of salt application (Highways Agency)*

	Rate of application
Pre-salting	10–20 g/m²
After frost, ice or snow	20–40 g/m²

Over the Christmas shut-down, security measures should be maintained at an adequate level, with emergency procedures in place. Dewatering or pumping operations need to be managed to avoid the pumps running out of fuel or malfunctioning, which could cause subsequent flooding of the site. Where discharges are continued throughout site closure, they will need to be checked and any problems reported and rectified.

12.5.6   Ecological constraints

> **Road salt and de-icing**
>
> - Direct drainage from stores of de-icing agents to the foul sewer, with prior written authorisation from the statutory sewerage undertaker.
> - Use up-to-date weather forecasts to avoid applying de-icing agents unnecessarily.
> - Keep records of all de-icer consumption.
> - To minimise the risk of affecting controlled waters when applying rock salt, do not exceed the recommended application rates (see Table 12.1). Porous wearing courses may require a higher rate.
> - Keep up-to-date with de-icer chemical development. Compounds may be available with a much lower BOD than the glycol equivalents. The use of urea is strongly discouraged, as ammonia, one of the breakdown products of urea, is highly polluting in watercourses and groundwater.
> - Review application techniques to minimise the amount of de-icing agent used.
> - If any part of the site is within a source protection zone or located on a vulnerable aquifer, consider taking additional precautions.

The timing of construction activities in relation to ecological constraints is particularly important. Sensitive ecological sites and protected species may have been identified in ecological surveys and the project environmental statement, where one exists. When programming the works (including scheduling additional ecological surveys) it is important to take account of critical times of the year (eg hibernation, breeding or nesting seasons).

> A contractor received a land drainage consent to install a potable water pipeline under a major river. Just before work on the six-week pipejack crossing started, it was noted that the consent prevented operations between October and April because of the fish-spawning season. This resulted in significant programme delays and additional costs.

Work programmes should also recognise there may be a need to obtain licences or permissions for works in areas of special designation (SSSI, SPA, cSAC etc). Additional restrictions imposed in the conditions of any licence or consent, such as working width, methodology, access routes, should also be recognised and passed on to the site team.

There are certain species, protected by law, that are often affected by changes to the water environment as a result of construction works. They include newts, water voles and fish. Chapter 24 provides guidance on ecological constraints; see also CIRIA C587 *Working with wildlife* (Newton et al, 2004a).

12.5.7 Water amenity

Additional restrictions on the timing of the works may be imposed where construction activities affect navigation or the general amenity value of a watercourse. Restrictions on construction works are often imposed to avoid peak seasons and bank holiday weekends. Significant works on inland waterways such as canals are often carried out only during winter months (November–March) when leisure use is less intense. Works taking place on watercourses near to coastal areas (and some distance inland) may also be restricted to avoid the summer (July–August) in order to maintain water quality around coastal waters and beaches. This is particularly important for beaches awarded blue flag status.

12.5.8 Reinstatement and landscaping

Early consideration needs to be given to landscaping and the reinstatement of land following completion of the works. A phased programme of reinstatement to allow for grass and plants to be established early, rather than at the end of the project, can help reduce the risk of water pollution from the project in its permanent state. Seeding is most successful when carried out during the growing season (March–October), while the planting of trees and shrubs is best undertaken during the dormant winter months (October–March). Interim measures such as erosion control matting and the maintenance of the site temporary works drainage can be used to reduce the risk of pollution until vegetation becomes established. For further information see Chapter 19.

12.6 ROLES AND RESPONSIBILITIES

It is important to define the environmental responsibilities of specific organisations and personnel within the project management structure. On linear projects, roles are often duplicated to cover several individual sections of the route. Certain responsibilities may exist in one section that are less relevant in other sections: for example, a settlement pond and outfall located in one section of the scheme receives site drainage from the whole site. Staff should be aware of their overall environmental responsibilities, rather than passing the responsibility solely to the relevant section.

> Where a scheme is managed in separate sections be aware that activities in your section of work may affect other sections, for example through site drainage.

Name and job role should be recorded on a project organisation chart and lines of communication including the interface with project staff and other key stakeholders clearly defined. The environmental responsibilities of key staff may be defined in a letter of appointment or identified in work procedures. Particular roles and responsibilities will be defined by project-specific requirements, but examples of typical roles are outlined below.

Site manager/project manager/site agent – has principal responsibility for environmental management on the project.

Site environment manager or **HS&E manager** – supports the site manager by co-ordinating the environmental aspects of the project, advising other site staff, monitoring the arrangements in place and providing feedback on project environmental performance. The environmental manager ensures, in particular, that any mitigation measures identified in the ES are implemented. Where a dedicated environmental manager is not on site, the health and safety manager may take on environmental duties in addition to health and safety supervision. Overall responsibility for environmental management nevertheless remains the responsibility of the site manager.

Works manager/site engineer – are often in the best position to put the environmental plan into practice. They need to understand the relevant environmental obligations on the scheme and the practical measures required to comply with them. On large projects engineers may be responsible for the implementation of the plan on particular sections of the work.

Site foremen/supervisors – work closely with subcontractors and site operatives and ensure that environmental controls are implemented at the workface.

In addition, there may be trained personnel to manage particular tasks such as refuelling plant and equipment, managing the stores, water quality monitoring and supervising the segregation and collection of waste.

12.6.1 Management of subcontractors

The environmental awareness of subcontractors varies considerably. Some may be very conscientious, while others may assume that all responsibility rests with the principal contractor on site. The site manager must ensure that the potential risks of water pollution (and other environmental risks), contract requirements and obligations are communicated to subcontractors as early as possible (ie at invitation to tender).

Main contractors should provide subcontractors with suitable areas for refuelling, material storage, washout etc during site set-up (see Chapter 16), and facilitate best practice.

Major contractors have found they gain benefits by working with their subcontractors to improve their environmental performance by arranging training for example.

Arrangements for the on-site management of nominated subcontractors (that is, those nominated by the client) and contractors carrying out enabling works in preparation for the main works should be no different to that of a traditional subcontract, with the principal contractor maintaining full control and responsibility.

Note that subcontractors are strictly liable for any pollution they cause, but the main contractor can also be prosecuted for knowingly allowing pollution (see Section 10.3).

Management of subcontractors

1. Select subcontractors on their competence and past performance
2. Communicate risks, contract requirements and obligations early (at tender)
3. Ensure that subcontractors personnel are aware of the risks before starting work (site induction and method statement briefings).
4. Monitor subcontractors performance to ensure that obligations are met and best practice is implemented.

12.6.2 Training

On-site training is part of the internal communication process and is essential in minimising the risk of water pollution. Site personnel need to be aware of particularly sensitive areas, high-risk activities and the implications of pollution and have a clear understanding of their roles and responsibilities. On-site training may take the form of site induction, method statement briefing or toolbox talks, posters and site notices. Arrangements for training should cover all site personnel including subcontractors.

Key guidance

Published training materials

CIRIA and EA (1996). CIRIA SP141V *Building a cleaner future*, video, poster, booklet

Murnane *et al* (2002). CIRIA SP156 *Control of water pollution from construction sites – guide to good practice*

Construction Confederation Environmental Forum (2005). *Environmental induction video and toolbox talks*, <www.thecc.org.uk>.

12.7 EMERGENCY PROCEDURES

It is essential that emergency and contingency planning form part of the environmental management arrangements on site. This should include a list of key emergency contacts and key roles and responsibilities in the event of a water pollution incident. All personnel need to be aware of the emergency arrangements on site. For further information see Chapter 15.

Site planning

1. Establish project plan and organisation – clearly identify roles and responsibilities.
2. Identify key receptors and assess potential risks.
3. Engage with regulators, statutory bodies and other key stakeholders.
4. Establish any constraints on programming.
5. Confirm water quality baseline and identify suitable control measures and monitoring.
6. Identify the need for training and awareness and communicate requirements to the workforce (including subcontractors).
7. Establish emergency procedures.

13 Licences and consents

The following activities generally require written permission from the regulatory bodies:

- discharging water to surface water or groundwater
- discharging sewage to ground or to a foul sewer
- abstracting or pumping water from surface water or groundwater
- working in or near water
- working in tidal waters.

Details on when, how and why to apply for consents are provided in the following sections. The "Application Forms and Guidance" section of the NetRegs website provides additional information and links to electronic forms, where available.

13.1 DISCHARGING TO SEWER

What do I need consent for?	To discharge to public foul sewer, you should either obtain written permission from the statutory sewerage undertaker in the form of a trade effluent consent or enter into a "trade effluent agreement" with the sewerage undertaker under the Water Industry Act 1991 (**England and Wales**), Water Industry (**Scotland**) Act 2002 and Water and Sewerage Services (**Northern Ireland**) Order 1973. A separate consent/agreement may be needed for discharges into different sewers along the route.
How to apply	In most cases, apply to your local water company (in **England** and **Wales**), Scottish Water (in **Scotland**) and the Water Service (in **Northern Ireland**). When discharging sewage effluent to anywhere other than a foul sewer (eg septic tank), apply to the EHS in **Northern Ireland** (<www.ehsni.gov.uk/pubs/publications/Appl_Consent_to_Discharge_Annex5.pdf>).
When to apply	The water undertaker has three months to decide whether to issue a consent.
Information needed	When applying for consent you will need to state the: • nature or composition of the effluent • maximum quantity proposed to be discharged on any one day • highest rate of discharge proposed.

Conditions	Consents may either be unconditional or subject to conditions covering:
	• the sewer or sewers into which effluent may be discharged
	• the nature and composition of the effluent:
	  ○ pH (typically pH 5–9)
	  ○ biochemical oxygen demand (BOD)
	  ○ chemical oxygen demand (COD)
	  ○ oxygen (mg/l)
	• the maximum quantity of effluent discharged in any one day (m^3/day)
	• maximum hourly flow (m^3/day)
	• the highest rate at which the effluent may be discharged
	• inclusion of other substances such as oils, chemicals
	• prohibition of certain substances.
	The absence of prohibitions or conditions on a consent does not mean all substances are permitted, and you will remain liable for any pollution.

Monitoring	Case-specific but monitoring of the following may be required to ensure compliance with a consent: • volume and rate • quality. Reducing the volume flow and monitoring can reduce the charges levied.

13.2 DISCHARGING TO SURFACE WATER OR GROUNDWATER

What do I need consent for?	The discharge of any matter to surface or groundwater in **England**, **Wales** and **Northern Ireland** requires a written "discharge consent" issued under the Water Resources Act 1991 (as amended by the Environment Act 1995 and the Water Act 2003) and the Water (Northern Ireland) Order 1999. If water is being abstracted prior to discharge, eg from an excavation or through dewatering to lower the water table, a transfer licence may be required. See Section 13.3 for more details on abstractions and dewatering.
	From April 2006, in **Scotland** building, engineering and other works in contact with groundwater, and discharges to surface and groundwater will require authorisation under the Water Environment (Controlled Activities) (Scotland) Regulations 2005. There are three tiers of authorisation: general binding rules (GBRs), which cover specified low-risk activities; registration, which applies to activities with a predictable risk; and licences for activities that need specific conditions to effect environmental protection. Licences are divided into a simple licence, in which standard conditions apply, and complex licences that require environmental assessment and site-specific conditions. Applications are required for registration and licences. Determination will be based on SEPA taking a risk-based approach so that controls are proportionate to the risk to the environment. Where authorisation is required for several activities at a single site in Scotland (discharge, dewatering, works in water), a single composite authorisation will be issued that covers all those activities.
How to apply	Apply to the Environment Agency in **England** and **Wales**, <www.environment-agency.gov.uk>, the Scottish Environment Protection Agency (SEPA), <www.sepa.org.uk>, in **Scotland**, or the Environment and Heritage Service in **Northern Ireland**, <www.ehsni.gov.uk/pubs/publications/Appl_Consent_to_Discharge_Annex2.pdf> for trade effluent and <www.ehsni.gov.uk/pubs/publications/Appl_Consent_to_Discharge_Annex5.pdf> for sewage effluent.
When to apply	On receipt of the application the Environment Agency and EHS have up to **four months** to decide whether to issue a consent in **England** and **Wales** and **Northern Ireland**.
	In **Scotland**, SEPA has **30 days** to determine an application for Registration and **four months** to issue a licence.
Information needed	This depends on type of application. Generally, when applying you will need to state the following information on the application form: • grid reference location of the discharge point • volume and rate of discharge • nature and composition of the discharge • receiving medium • type of treatment prior to discharge • monitoring proposed.
Conditions	Consents/licences are subject to certain conditions, depending on the nature of the discharge and the quality of the receiving waters, and may cover: • grid reference location of the discharge point • pH • levels of suspended solids, hydrocarbons and possibly other substances depending on source (note restrictions on List I and List II substances in the Groundwater Directive) • temperature • maximum volume and flow rate. The absence of prohibitions or conditions on a consent/licence does not mean all substances are permitted, and you will remain liable for any pollution.
Monitoring	Case-specific, but usually require the following to ensure compliance with a consent: • volume and rate • quality • procedures for monitoring outfall, recording and maintaining records.
Additional information	Landowner's agreement, including the riparian owner (eg British Waterways), is also required.

13.3 ABSTRACTING AND DEWATERING

What do I need consent for?	Abstraction (including dewatering) in **England** and **Wales** of more than 20 m³ per day from surface water or groundwater requires one of three forms of licence under the Water Act 2003.
	
	1. A *temporary licence* is required for any abstraction lasting less than 28 days.
	2. A *transfer licence* is required for abstraction of water for 28 days or more from one source of supply to another without intervening use – for example, transferring water from an excavation to a watercourse, dewatering to a watercourse to lower groundwater, or overpumping from one watercourse to another.
	3. A *full licence* is required for any other abstraction for 28 days or more.
	
	Note (England and Wales): If you are dewatering or pumping water that has collected in an excavation or shaft then you will not require an abstraction licence. If you intend using water from pumping or dewatering operations for dust suppression on site, pressure testing etc, you may then require an abstraction licence. You may also need a discharge consent to dispose of or transfer dewatered water (see Section 13.2).
	
	In **Scotland**, there are three levels of authorisation under the Water Environment (Controlled Activities) (Scotland) Regulations 2005. SEPA will carry out a risk assessment on receipt of an application to determine which authorisation is appropriate. Where authorisation is required for several activities at a single site (discharge, dewatering, works in water), a single composite authorisation may be issued in respect of those activities.
	
	1. General binding rules (GBRs) – these rules cover specified low-risk activities and are effectively seen as good practice. Operators whose activities meet GBRs do not need to contact SEPA:
	   - abstractions less than 10 m³/day are covered by a GBR.
	2. Registration – applies to activities with a predictable risk; operators need to apply to SEPA:
	   - abstractions of between 10–50 m³/day.
	3. Licensing – applies to activities that require specific conditions to afford environmental protection. There are simple licences where standard conditions will apply and complex licences, which will entail environmental assessment and specific conditions:
	   - simple licence for abstractions 50–100 m³/day
	   - complex licence for abstractions of > 100 m³/day.
	
	Under the Water (**Northern Ireland**) Order 1999 there exist powers to make regulations for the control of water abstraction. You may need to obtain permission to abstract water from certain bodies of water, for instance when the source affects a protected habitat area.

How to apply	Apply to the Environment Agency in **England** and **Wales**. (Note: at the time of writing this process was the subject of government consultation; contact the Environment Agency for information.) Contact the Scottish Environment Protection Agency (SEPA) in **Scotland** or the Environment and Heritage Service in **Northern Ireland** for further information.
When to apply	On receipt of the application the Environment Agency has up to three months to decide whether to issue a licence in **England** and **Wales**. In **Scotland**, SEPA has 30 days to determine an application for registration and four months to issue a licence.
Information needed	Depends on type of application. Generally, when applying you will need to state the: - volume - source and use - other information as may be required.
Conditions	Consents are usually subject to certain conditions including: - volume - grid reference location of the discharge point.
Monitoring	Case-specific, but likely to include volume and rate.

13.4 WORKS IN OR NEAR WATER (including outfall and bank-side structures; see Section 13.5 for tidal waters)

What do I need consent for?	In **England** and **Wales** a flood defence consent is required under the Water Resources Act 1991 and the Land Drainage Act 1991 for:
	
	• works in, over or under any main river (identified by the Environment Agency, Defra and Welsh Assembly Government)
	• works in, over or under all other watercourses (ordinary watercourses) if the flow is likely to be affected
	• temporary works affecting the channel of main rivers or ordinary watercourses
	• temporary and permanent works in the floodplain of main rivers (a floodplain may extend some considerable distance from a watercourse).
	
	Works within 7–10 m from the top of the bank may also require consent. Confirm with the local Environment Agency to determine whether consent is needed. "Works" include temporary works, for example, haul road, culvert, stream diversion; as well as permanent works, eg new road bridge, underground pipeline. Where a route requires numerous water crossings, for example, a long pipeline, then a single consent can apply to several locations.
	
	In **Scotland** building, engineering and other works in or near inland surface water and wetlands, as well as works in contact with groundwater, require authorisation under the Water Environment (Controlled Activities) (Scotland) Regulations 2005. There are three tiers of authorisation: *general binding rules* (GBRs), which cover specified low-risk activities; *registration*, which applies to activities with a predictable risk; and *licences* for activities that need specific conditions to effect environmental protection. Licences are divided into a *simple licences*, where standard conditions apply, and *complex licences*, which require environmental assessment and impose site-specific conditions. Applications are required for registration and licences. Determination will be based on SEPA taking a risk-based approach so that controls are proportionate to the risk to the environment. Where authorisation is required for several activities at a single site (discharge, dewatering, works in water), a single composite authorisation will be issued to cover all those activities.
	
	The following examples show the tier of control likely for specified activities. SEPA has the discretion to require an operator to apply for registration or a licence if it believes that pursuance of an activity under GBR or registration will be insufficient to ensure adequate environmental protection.
	
	1 *General binding rules* (GBRs) – these rules cover specified low-risk activities and are effectively seen as good practice. Operators whose activities meet GBRs do not need to contact SEPA. GBRs may apply to:
	   • pipeline/cable crossing by boring
	   • temporary bridges or culverts
	   • minor bridges
	   • small-scale bank reinforcement.
	
	2 *Registration* applies to activities with a predictable risk. Operators need to provide basic information to SEPA by way of an application. Registration may be required for:
	   • croys
	   • boulder placement for fisheries enhancement
	   • bridging culverts
	   • pipeline crossing by cutting
	   • medium-scale unconsolidated bank reinforcement/bank reprofiling
	   • gravel extraction from dry gravel beds of watercourses.
	
	3 *Licence – simple and complex.* A licence is likely to be needed for:
	   • culverts
	   • canalisation
	   • permanent diversion
	   • flood protection works
	   • pipeline/cable crossing by direct cutting
	   • large-scale resectioning/dredging
	   • solid bank reinforcement
	   • bridges (involving construction in channel).

What do I need consent for? (contd)	In **Northern Ireland**, if you intend to interfere with the flow of water on a watercourse – by culverting water, for example – you should consult the Rivers Agency. It is an offence under the Water (Northern Ireland) Order 1999 to deposit into a watercourse any polluting matter or anything that could impede the water's proper flow. Consultation with both the Environment and Heritage Service (EHS) and the Rivers Agency is advisable. **Note:** a discharge consent may **also** be required if any water pumping operations are planned. **Note:** for works in the tidal zone a FEPA licence may be required (see Section 13.5 below).
How to apply	Apply to the Environment Agency in England and Wales. Apply for authorisation to the Scottish Environment Protection Agency (SEPA) in Scotland. Discuss issues in Northern Ireland with the EHS and the Rivers Agency.
When to apply	On receipt of the application the Environment Agency has up to two months to decide whether to issue a consent in England and Wales. SEPA has **30 days to determine an application for registration** and **four months** to issue a licence in **Scotland**.
Information needed	Depends on type of application. Generally, when applying you will need to state the: • location • method of work • measures to prevent pollution.
Conditions	Consents/licences are usually subject to certain conditions including: • agreed methodology • timing • reinstatement.
Monitoring	Monitoring requirement depends on the type of work and the sensitivity of the watercourse. You may need to undertake pre-construction and post-construction water quality monitoring. Ecology surveys may be required during the appropriate season before work begins. If certain species are involved, the surveys must be conducted by licensed ecologists (see Chapter 24).
Additional information	Works affecting British Waterways-owned canals must follow its code of practice (available at <www.britishwaterways.co.uk/images/COP_2005.pdf>). The inland drainage board should be consulted regarding works in or near watercourses where it has responsibility. A development licence from the conservation body may also be required if the works are likely to disturb certain species or habitats (see Chapter 24).

13.5 WORKS IN TIDAL WATERS

What do I need consent for?	In the UK construction licences are required for the placement of materials in the tidal zone below mean high water springs (MHWS), which includes the tidal waters of any estuary, creek, bay or river, under the Food and Environment Protection Act 1985 (FEPA). Construction is taken to include a wide range of activities in this context, including land reclamation, ground investigation, bridges and bridge piers, pipelines and outfalls, dredging, tunnels, overhead cables and flood defences. Activities such as laying cables, pipelines or outfalls in or under tidal waters may also require consent under the Coast Protection Act 1949 or the Telecommunications Act 1984. As the circumstances in these situations vary considerably, the project promoter or contractor should seek advice on the consents required.
How to apply	In **England** and **Wales** apply to the Marine Consents and Environment Unit (MCEU), which processes consents on behalf of Defra and the Welsh Assembly Government. Application forms and guidelines can be downloaded from <www.mceu.gov.uk/mceu_local/fepa/fepa-start.htm>. In **Scotland** apply to the Fisheries Research Services (FRS), <www.marlab.ac.uk>, an agency of the Scottish Executive Environment and Rural Affairs Department. In **Northern Ireland** apply to the Environment and Heritage Service (EHS), <www.ehsni.gov.uk>, an agency within the Department of the Environment. Forms can be downloaded from <www.ehsni.gov.uk/word/FEPA/application%20forms/ConsAppl.pdf>. The MCEU, FRS and EHS similarly process consents for marine works under the Coast Protection Act (or equivalent local harbours legislation) and the Telecommunications Act where appropriate.
When to apply	A period of at least 10 weeks is normally needed to assess and process an application in **England** and **Wales** and **Scotland**. The EHS requests application four months in advance of works in **Northern Ireland**. However, some projects may raise matters that take a significantly longer to assess and for any problems to be resolved. Licences generally last 12 months; where works will take longer, consents should be renewed 10 weeks before they expire.
Information needed	When applying you will need to state the: - type of consent application - location, duration and description of project - construction drawings and sections - detailed methodology of works and measures to be taken to minimise risks to the marine/water environment - description of materials to be used - details regarding protected conservation sites.
Conditions	Licences may be subject to specific conditions, which typically include: - specific method of working - controls on material deposition - timing and tidal state for dredging and/or disposal - water quality monitoring - ecological surveys - lighting of works for navigation purposes.
Monitoring	Likely to include baseline water quality monitoring and monitoring during the works, particularly for suspended sediments.

14 Monitoring

Monitoring is an essential tool for establishing and observing the status of the water environment along and adjacent to the site. It can demonstrate whether or not construction activities are having a detrimental impact on water quality and quantity, and thereby allows action to be taken to protect the water environment. This is particularly important where sensitive on- or off-site environments exist, such as licensed abstractions for private, industrial, commercial use and sites of ecological importance. Monitoring can also identify where pollution is coming from an off-site or upstream source.

14.1 LEGAL REQUIREMENTS

Any consents and permissions issued by the environmental regulator or sewerage undertakers are likely to include conditions for monitoring the quantity, quality and flow rate of water. The results must be made available to the environmental regulator or sewerage undertakers on request.

> It is illegal to breach the conditions of discharge consent. Contractors should be aware that the named consent-holder is responsible for compliance with consent. Therefore, they would be prosecuted if a subcontractor or other third party caused breach of the consent conditions (see Section 10.3).

Where water quality issues are particularly significant on a scheme, it is likely that any ES mitigating measures, any planning consent, or indeed the contract documents will include conditions to protect the water environment, involving water monitoring, to be in place on site. These documents, and the requirements therein, are legally binding.

14.2 BENEFITS OF MONITORING

The benefits of water monitoring and maintaining monitoring records include:

- understanding the existing quality of the water environment before the start of construction work to be able to compare it with water quality during and after construction
- early warning of declining trends in water quality
- identifying off-site sources of pollution and understanding the potential for movement of pollution towards the site
- controlling the effects of dewatering works on water levels
- discharge consent compliance
- compliance with the abstraction licence
- evidence of "best practice" procedures
- supporting evidence in the event of a pollution incident.

An adequate monitoring programme can highlight potential problems before they arise, allowing time to implement mitigation measures, avoiding costly prosecutions, clean-up and compensation.

14.3 WHAT TO MONITOR

Monitoring of water features can be carried out for one or more of the following:

- physical character of water feature – size, location etc
- flow/discharge characteristics – water levels, flow volume (of outfall or pipe) etc. Note that dry weather low flows can vary considerably from wet weather peak flows
- water quality – chemical analysis for a range of parameters such as pH, chloride, suspended solids, nitrate, BOD, dissolved oxygen, hydrocarbons and metals (lead, zinc, cadmium etc)
- ecological status – specialist surveys for habitat type and aquatic and terrestrial plants, invertebrates, fish, mammals etc.

A risk-based approach should be used when determining what and where to monitor, based on the construction activities, the length of the project and the sensitivity of the location (see Table 14.1 and Section 12.3). The degree of monitoring should be commensurate with the level of risk identified, thus allowing better use of resources. A monitoring regime should be discussed and agreed with the environmental regulator. Any changes to the regime should be detailed and justified.

Table 14.1 *Example water monitoring requirements*

Construction activity	Site sensitivity	Monitor for*
Construction near watercourse	Watercourse not designated or a water supply, fishery downstream	Suspended solids, oil and pH (monitors for silt, oil, fuel and concrete)
Surface water discharge into watercourse	Watercourse not protected; discharge licence in place	Monitor outfall to meet conditions of the discharge consent
Works near watercourse	Watercourse designated as site of special scientific interest	Range of chemical parameters in agreement with conservation body
Dewatering	No groundwater aquifer, watercourse nearby	Water levels in site boreholes; water levels in watercourse (particularly in summer)
Dewatering	Aquifer supplying drinking water	Water levels in boreholes and samples tested for a range of parameters

* Monitoring requirements will be specific to individual sites and should be agreed with the environmental regulator

14.4 WHEN TO MONITOR

14.4.1 Pre-construction surveys

The baseline (pre-construction) conditions need to be determined for each water feature at and adjacent to (within 500 m is generally adequate, although outfalls or other features may be farther away) the site (see Chapter 2 and Section 8.2). Where appropriate it will be necessary to include artificial water features (eg surface water outfalls), groundwater and also ephemeral ditches and field drains, which may only flow during wetter conditions. Detailed surveys are usually undertaken by specialists, and may have to be undertaken in accordance with the ES, planning or contract documentation (see Chapter 7).

Where detailed surveys are not required in the ES, planning or contract documentation, it is always worth carrying out baseline pre-construction monitoring of water quality.

Pre-construction monitoring can be undertaken along the route as a whole, far in advance of works (for example during an EIA), and over a long period of time (eg groundwater monitoring), but is also highly beneficial immediately prior to a particular construction activity such as a watercourse crossing, where monitoring could simply be for suspended solids and hydrocarbons.

14.4.2 During construction

Monitoring during construction can encompass water quality, water level and/or ecology, depending on the sensitivity and location of the route or part of the route. If continuous monitoring is required, samples should be taken from the same locations as the pre-construction surveys so that results may be compared.

Even where monitoring is not required by contract documentation, it is always worth carrying out some monitoring of the works, perhaps by visual inspection or limited water sampling of particular activities. See Section 18.4 for suggested techniques.

In addition, sampling and maintaining records to monitor compliance with discharge or abstraction consents is usually a condition of such consents.

14.4.3 Post-construction surveys

Where the site runs across, or is near to, a sensitive environment, the contract documentation may require post-construction surveys that can be compared with the pre-construction surveys. These are used to:

- demonstrate that there has been an environmental improvement
- demonstrate that there has been no impact, and the environment has remained in, or returned to its pre-construction status
- to identify where further mitigation may be required to improve the water environment to its status before construction.

Where water quality has been monitored before a particular activity has taken place, eg construction of a watercourse crossing, similar monitoring immediately after the works are completed is very worthwhile to demonstrate whether the working methods were appropriate in reducing pollution, and the timeframe over which any impact was experienced. In the event of a pollution incident occurring as a result of similar works elsewhere, the results of such monitoring can demonstrate that best practice has generally been successfully implemented on site.

14.4.4 Post-pollution incident monitoring

In the event of a pollution incident occurring, water quality should be monitored at and downstream of the incident. The results can be used to assess the level and geographical extent of pollution that has occurred, the impact it may have had, and the timeframe over which the impact was experienced.

Ideally, any potential risks should have been identified on a risk assessment and an emergency plan put in place (Chapters 12 and 15), which will include monitoring procedures.

14.5 HOW TO MONITOR

There are various on-site monitoring techniques available to monitor both surface water and groundwater. The recommended method depends on the sensitivity of the water body, the key characteristics that are most at risk (eg water level, sediment concentration, fish count), any specified range of parameters to be monitored and the level of accuracy required.

Approach the environmental regulator with proposed monitoring methods and an initial proposed monitoring regime, and seek their advice and agreement.

> **Key guidance**
>
> BS 5667-6:2005 *Water quality. Sampling. Guidance on sampling of rivers and streams*
> BS 6068-6.11:1993 *Water quality. Sampling. Guidance on sampling of groundwaters*
> Preene *et al* (2000). CIRIA C515 *Groundwater control – design and practice*
> Newton *et al* (2005). CIRIA C587 *Working with wildlife*

14.5.1 Surface water

> Always undertake a risk assessment prior to staff working in or near water. Follow health and safety guidance such as that at <www.healthandsafety.co.uk/water>.

Visual inspection

The simplest monitoring technique is visual inspection, which is inexpensive and gives immediate results. Any competent member of on-site staff can undertake visual monitoring using an inspection procedure (see checklist Figure 14.1).

Any changes, such as discoloration, odour, oily sheen or litter, should be noted (see On-site tests, below, for turbidity tests). Water bodies, outfalls or other receptors ought to be inspected at regular intervals, depending on the activities taking place. High-risk activities and locations, such as settlement tank outfalls from dewatering below-ground excavations, should be inspected at least daily. Also check that any equipment (such as a flow meter) is operating correctly, and whether materials such as straw bales or oil-absorbent materials need replacing.

> There is a type of harmless bacteria that can form a multicoloured sheen on the surface of standing water. This sheen looks similar to that caused by oil on the surface of the water. The easiest way of distinguishing the two is to drag a stick through the material. If oil is present, then it will usually remain as a constant layer. If the sheen is caused by bacterial growth, then it will break up into smaller pieces with obvious and irregular edges. If in doubt, you should have the water tested.

Visual inspection of the water level in surface water bodies is also simple method giving immediate results. Compare water levels to a known, fixed point, or use a survey pole to get more robust results.

Dust can also pollute water bodies if it settles on the water surface. Visual assessment of dust on site is the most effective way of monitoring dust generation and likely water pollution risk. (There are more detailed methods by which the public nuisance and health and safety hazards posed by dust can be monitored; see BS 1747-1 to 12:1969 to 1993 for more information.)

Keep a record of the findings of all visual inspections (see Section 14.6 below) and make sure any identified actions are carried out. While records of visual inspections are very useful, they are not adequate for demonstrating compliance with discharge or abstraction consents, but they can be used to supplement weekly or fortnightly sampling, and to demonstrate best practice.

> Where pollution is observed, report it according to the established site procedures (Chapter 15) and take immediate action.

On-site tests

On-site kits are available for basic chemical tests providing almost immediate results. Typical kits include litmus papers (for pH), flow meters, oxygen meters and portable sample kits with reagents for a range of substances. Turbidity tubes or turbidity meters are very simple test, frequently used to monitor suspended solids (turbidity) as well as a range of other substances following the addition of reagents. Turbidity is measured in NTUs (nephelometric turbidity units). Users should follow the suppliers' advice and the instructions accompanying the equipment.

Laboratory analysis

The most costly, but also most accurate, method of monitoring is sample collection and laboratory analysis. Any competent member of staff following a sampling procedure can collect samples, but the laboratory should always be contacted beforehand, so it can provide appropriate containers and advise on the required tests, storage and transport. For detailed or long-term monitoring it is advisable to use specialists, as the sampling methods, scheduling of laboratory tests and interpretation of the results require expert input.

The detection level of the specified test should be suitable for the accuracy of the results required. Laboratories can use different test methods to obtain greater levels of accuracy. The volume of the sample can also affect the detection limit (more samples will give a lower detection limit).

Laboratory analysis is suitable for demonstrating discharge consent compliance. The Environment Agency requires all laboratory testing to be UKAS-accredited for the tests specified.

> Laboratory analysis and the accuracy of the results depend on samples being taken using appropriate and accurate techniques.
>
> The following factors are critical:
> - sampling technique (location and depth, sampling equipment, cleaning with distilled water)
> - use of pre-prepared bottles (washed with or containing certain reagents)
> - required volume of sample (can range from 100 ml to 2 litres)
> - addition of certain reagents at time of sampling
> - storage and transportation temperature
> - accompanying documentation (chain of custody).

14.5.2 Groundwater

The need for groundwater monitoring (see Chapter 14) depends on the construction activities and the sensitivity of the groundwater. Monitoring of groundwater levels is likely to be required for dewatering from or near an aquifer, if consented in the first place (see Chapter 13). Major excavations in or near contaminated land are likely to require monitoring of groundwater quality to assess any leaching or migration of contamination etc. Any groundwater monitoring programme should be agreed with the environmental regulator. Refer to the publications in "Key guidance" above for further detailed information.

Groundwater levels in conventional, permanent monitoring boreholes can be monitored by installing standpipes or piezometers at specific depths along the route. Ideally, there should be a year's pre-construction water level data where the borehole has been dipped every 2–4 weeks, as groundwater levels vary with season. Groundwater-fed springs should be identified and inspected, water quality monitored if required (based on likely risk of pollution), and flow volumes inspected or monitored to detect any unnatural changes.

Although groundwater samples can be obtained using conventional pumping methods and tested using on-site test kits, the groundwater quality standards are such that laboratory analysis (see above) is usually more appropriate.

14.5.3 Ecology

Ecological surveys are generally carried out before construction work begins, prior to particular activities taking place and then following completion of construction. Surveys are usually either for habitat type and quality, for a range of species (eg aquatic invertebrates), or for a particular plant or animal species (water voles, for example). As the banks of watercourses form part of the water habitat, surveys of terrestrial species should also be included. Guidance on ecology is provided in Chapter 24.

Ecological surveys can only be undertaken at certain times of the year depending on the form of survey. Qualified ecologists must be appointed to carry out surveys and licences may be required for certain species. Sufficient time should be allowed for licences to be obtained and surveys carried out. See Appendix 3 for guidance on survey timing.

14.6 RECORDS

It is essential to keep a written record of all monitoring undertaken. For visual inspections or on-site tests, it is particularly useful to record observations on a proforma sheet (or transfer the information from a site notebook) so the records can be maintained on file. This ensures the records are traceable when staff change and are readily accessible to site management. Figure 14.1 shows an example form used to record checks on river water quality near site. Similar forms could be used for discharge quality checks etc.

When taking samples for analysis, visual observations should be recorded, as they can help in the interpretation of the results.

> **Monitoring records should include:**
> - staff member
> - date and time
> - weather conditions prior to and at time of monitoring
> - monitoring results or observations
> - construction activities (where appropriate)
> - actions required (where necessary).

Monitoring records can be used to demonstrate compliance with discharge consents or abstraction licences. In the event of a pollution incident, monitoring results can indicate whether the water environment was affected and can provide evidence that the contractor has been acting responsibly in protecting the water environment.

If the site manager or environmental manager is in regular contact with the environmental regulator or nature conservation body, the authorities may wish to receive copies of the results as reassurance that the site is operating responsibly. Under certain contracts, regular formal reports on monitoring results are required to be issued to the client and/or statutory environmental authorities. Project extranets are increasingly being used to make this information available to the environmental regulators.

> **Monitoring**
>
> 1. Review the contract documentation (including planning consent, EA etc) to identify any requirements for monitoring or surveys of surface water, groundwater or aquatic ecology.
> 2. Assess the risks of the proposed construction activities to the water environment.
> 3. Produce a proposed monitoring programme taking into account any timing restrictions.
> 4. Agree the programme with the environmental regulator.
> 5. Undertake baseline pre-construction monitoring of water quality, flow, water levels and/or ecology according to agreed programme.
> 6. Implement a programme of routine monitoring or inspection of critical areas (eg outfalls, at watercourse crossings, in boreholes) and compliance points (eg consented discharges).
> 7. Maintain records of all visual inspections, monitoring and laboratory analysis.
> 8. Review the monitoring results and take action where necessary to improve water quality, repair equipment and replace water treatment measures (oil booms etc).
> 9. Implement post-construction monitoring, where required.

Insert contract name							Contract no		Insert watercourse name	
Date	Time	Site description	Visual check (circle answer)				Turbidity reading (NTU)	Comments	Monitor's initials
			Site activity	Water clarity	Oil film	Rainfall last 24 h			
			Low Med High	Clear Cloudy Coloured	None Minor Significant	Low Med High			
			Low Med High	Clear Cloudy Coloured	None Minor Significant	Low Med High			
			Low Med High	Clear Cloudy Coloured	None Minor Significant	Low Med High			
			Low Med High	Clear Cloudy Coloured	None Minor Significant	Low Med High			
			Low Med High	Clear Cloudy Coloured	None Minor Significant	Low Med High			
			Low Med High	Clear Cloudy Coloured	None Minor Significant	Low Med High			
			Low Med High	Clear Cloudy Coloured	None Minor Significant	Low Med High			
			Low Med High	Clear Cloudy Coloured	None Minor Significant	Low Med High			
			Low Med High	Clear Cloudy Coloured	None Minor Significant	Low Med High			
			Low Med High	Clear Cloudy Coloured	None Minor Significant	Low Med High			
			Low Med High	Clear Cloudy Coloured	None Minor Significant	Low Med High			
			Low Med High	Clear Cloudy Coloured	None Minor Significant	Low Med High			

Figure 14.1 *Example proforma for visual inspections of river*

15 Emergency and contingency planning

A contingency plan for pollution incidents or emergencies is an essential element of construction project management. An emergency plan should be developed during the project design phase and reviewed regularly so that it continues to apply to current construction activities. The emergency plan may be incorporated into an environmental management plan for the site.

> **Key guidance**
>
> EA, SEPA and EHS (2004a). PPG 21 *Pollution incident response planning*
>
> Fax or email forecasts from the Met Office's MetBuild Direct service, <www.met-office.gov.uk/construction/mbdirect/index.html>
>
> Emergency flood services and automatic telephone updates (where available) in the event of severe weather:
> <www.environment-agency.gov.uk/floodline>
> <www.sepa.org.uk/flooding/index.html>
> <www.riversagencyni.gov.uk/rivers/floodemergency-whotocontact.htm>.

15.1 RISK ASSESSMENT

The best way to manage pollution incidents is to prevent them. Incidents and emergencies are difficult to define and will depend on the nature of the site activities and the sensitivity of the water environments along the route. All possible risks need to be assessed and measures put in place to control them (Table 15.1). Emergency procedures should be developed and communicated to everyone on site.

Table 15.1 *Example measures to reduce pollution incidents*

Example pollution incidents and risks	Measures to reduce risk
Planned construction such as a bridge or a pipeline crossing a watercourse	Consult with regulatory bodies. Select an appropriate method of work to reduce impacts on watercourse (Chapter 20). Use of specially designed temporary works or plant is likely. Train staff
Unplanned incident such as bank collapse	Have a contingency plan in place to contain silt and other materials. Consider how materials and plant will be recovered
Unforeseen incident such as heavy rainfall or flooding	Review forecasts, rainfall and river data (see Chapter 18 and Key Guidance above). Have a contingency plan in place to direct and contain runoff from a 1 in 10-year event if possible. Install spillways to release water from settlement ponds in the event of exceptional rainfall (eg 1 in 50-year event)
Exceedance of discharge consent limit	Monitor regularly and take action if a trend in declining water quality is obvious (Chapter 14). Check operation of water treatment facilities regularly; maintain them – empty settlement tanks, replace oil booms routinely
Minor spills during refuelling or leaks from plant	Keep drip trays in place and spill materials available (Chapter 16). Train staff
Fuel spill as result of vandalism	Provide adequate site security; ensure fuel storage areas are secure and that containers are locked when not in use (Chapter 16)

15.2 EMERGENCY PLANS AND PROCEDURES

A simple spill response procedure (see Figure 15.1 for an example) should be displayed at appropriate locations on the whole site, at river crossings, near outfalls etc, and should state the following as a minimum.

1   Instruction to stop work and to switch off sources of ignition.

2   Contain the spill; location of spill clean-up material.

3   Name and contact details of responsible staff (these staff should assess the scale of the incident to determine whether the environmental regulator needs to be called).

4   Measures particular to that location or activity (for example, close pond outlet valve).

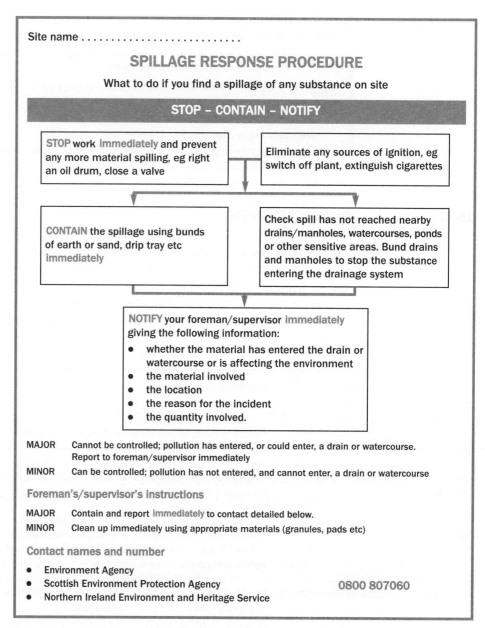

Figure 15.1   *Example emergency response procedure (Murnane et al, 2002)*

More detailed plans may be project-specific, location-specific or specific to a particular activity depending on the nature of the work. They should include details of site drainage, outfalls and watercourses to indicate where pollution may end up so that containment measures can be put in place at these locations.

At the start of the work contact should be made with the environmental regulator and other authorities. They are then more likely to understand the project, to be aware of and provide guidance on the measures being taken to address potential pollution and to provide advice when there is an emergency. The emergency services (particularly fire service) should be consulted to establish safe and appropriate access points and haul routes to site compounds and other areas where there may be a risk of spillage etc (eg outfalls, fuel storage).

The UK Spill Association (UKSpill) is a national trade association (replacing BOSCA) recognised by the environmental regulators representing the interests of the UK oil spill industry (<www.ukspill.org>). It can provide advice and contacts for suitable spill clean-up products and specialist contractors.

In an emergency, knowing the relevant people to contact for help can save time and minimise the impacts. To cover the full length of a route, more than one contact may be needed, so the plan should indicate which contacts apply to which sections of the site.

Obtain numbers for the following:

- radio/mobile contacts for site management and trained staff
- out-of-hours contacts
- environmental regulators (hotline 0800 807060 or local contact)
- water company (for spills to foul sewer)
- drainage authority (for some inland waterways not managed by the environmental regulators)
- riparian owner (for many inland waterways this will be British Waterways)
- downstream sensitive users (eg fish farm)
- spill response and clean-up contractors (use contractors who are members of UKSpill – see <www.ukspill.org/n_members.asp>).

15.3 TRAINING AND TESTING

Staff responsible for action in an emergency need to know their responsibilities. An incident at one location on the route may affect other parts of the site downstream, so it is important that someone be responsible for informing them. Train staff to use the necessary equipment such as spill kits or outlet valves.

Emergency arrangements need to be reviewed and tested periodically (and always after an incident) to ensure that measures are effective and that the workforce is aware of what to do in the event of an incident. Emergency drills should be recorded and improvements noted and actioned accordingly.

> **Reviewing the emergency plan**
>
> - monitor water quality and compliance with licences and consents (Chapters 13 and 14)
> - review number of pollution incidents
> - check measures are in place and functioning adequately, for example, settlement tanks, drip trays, cut-off drains
> - review staff training and awareness – ask a member of site staff where the nearest spill kit is, for example
> - check the spill response procedure is displayed at appropriate locations
> - check contact details are relevant and up to date
> - test the system with an incident drill.

15.4 EQUIPMENT

Emergency spill kits are available for dealing with spillage on both land and water. These should be obtained from a reputable supplier (UKSPILL-accredited) and used correctly. Other materials can be used to contain and treat emergencies involving solid materials and silt. Material safety data sheets and COSSH assessments provide advice on appropriate spill measures for chemicals, paints and hazardous materials.

The type of equipment required will depend on the project and the activities taking place. Table 15.2 shows the types and applications of emergency equipment. All site personnel should be trained and competent in their use.

Table 15.2 *Emergency equipment*

Equipment	Usage and limitations
Oil-absorbent granules ("cat litter")	Use only on LAND. Oil floats on water and the granules sink, so they will not work.
Organic oil absorbents	Commercial products similar to sawdust that absorb hydrocarbons on LAND and WATER.
Floating booms	Tubes of material intended either for containment of solid or immiscible liquids or for absorbing hydrocarbons on WATER. Make sure they are fixed closely up river banks or working platforms, are able to move with changing water levels and are accessible for clean-up and recovery.
Absorbent pads	Ideal for use in WATER, including drip trays and bunds that can fill with rainwater; can also be used on LAND. Absorbs hydrocarbons – diesel, petrol, oil-based paints, solvents and other organic liquids, eg chlorinated solvents, alcohol, antifreeze etc. To dispose of oily water from drip trays, pour the water through an absorbent mat before disposing of the "clean" water. Dispose of used mats in the appropriate place for oily waste.
Straw bales	Silt and coarser solids can be filtered from flowing WATER by using straw bales. This has its limitations in that the water flow must not be too fast, and must not be allowed to flow around, below or over the bales. Not suitable for clay or fine silts. Can be used if high levels of sediment are mobilised accidentally in WATER or to control unexpected runoff on LAND. Take care when removing the bales as the trapped silt can easily be remobilised.
Geotextile sheeting	Can be used either to wrap straw bales to filter out silts, to line a temporary straw bale settlement facility or to form a silt fence for runoff control on exposed ground (described in Chapters 18 and 19).
Timber planks	Useful as a skimmer board in ditches with slow-flowing water (traps floating materials and oils/fuel).
Ropes or wooden stakes	For fixing booms on either side of watercourses or around working areas (see "Floating booms" above).
Drain covers	Specially designed covers that form a watertight seal when placed over storm drains.
Putty	Used to seal damaged drums.
PPE	Oil-resistant gloves for use in recovering and disposing of used absorbent materials, overalls, masks.
Polythene sacks	These are often labelled as part of a kit for disposal of used oily clean-up materials etc. Double-bag and/or store in impermeable, bunded container prior to disposal as "hazardous waste".

Note: oil-dispersing chemicals are not recommended by environmental regulators as they increase the bioavailability of any spilled hydrocarbons.

15.5 CORRECTIVE ACTION

When an accident happens, it is important to learn from it and ensure that such an incident does not occur again. This may involve changing the method of work for a particular activity, providing containment or treatment materials, or simply training staff so they are aware of the correct method of work. Similarly if an audit of planned arrangements indicates that measures are not in place, or those in place need to be improved, action should be taken immediately.

A record of corrective actions and lessons learned should be kept. Where work areas on large projects are under different site management, the lessons learned need to be communicated across the project. Project-wide reporting (for example, incidents per month) is often successful in generating competition and motivation between work areas, in turn improving performance.

Checklist for action

- Prepare an emergency response plan for the project or for each work location.
- Assess the pollution risks and develop emergency and spill response procedures for site activities.
- Obtain details of key people you may need to contact for help.
- Provide equipment for dealing with pollution incidents.
- Train staff to follow procedures and use equipment correctly.
- Audit the emergency plan.
- Take action following an incident to ensure it does not occur again.

16 Site set-up

16.1 INTRODUCTION

A good site set-up is critical to the success of a project. In addition to a main compound, linear schemes, particularly large-scale projects, often require several smaller satellite compounds along the route to help service the works and provide welfare facilities for site personnel.

The set-up of a site may need to be considered long before the contractor is mobilised. Early contractor input is essential if a project is to be planned effectively for construction. The land take additional to that required for the route alignment will need to be known. If an environment impact assessment (EIA) is required, the location of site compound(s) will have to be determined and assessed.

> Once the EIA, planning consent or compulsory purchase has been carried out it is difficult to include compound and storage areas retrospectively. It is important therefore to consult the contractor at an early stage on the location, layout and size of the site compound, storage and lay down areas.

Factors that influence the location of a site compound include access, proximity to route alignment, land take, environmental sensitivity and the prior existence of facilities that could be used (eg power supply, hardstanding, water supply). A site risk assessment should be carried out to determine the best location for compounds (Chapter 12) and to identify any emergency or contingency measures that might be needed (Chapter 15). Early planning for the storage of potentially polluting materials, for supply and disposal of water, and for controlling runoff will reduce the risks of water pollution on site. There will also be a need to:

- locate the compound away from watercourses (including ditches) and aquifers
- avoid locations that are designated conservation areas
- identify areas with permitted access by public main road (reducing the need for haul roads)
- identify locations that have services in place (eg hardstanding, water supply, power and connection to foul drainage system).

Once the location of the compound(s) has been determined, consideration can be given to the type of equipment and materials to be stored and the layout of the compound(s). This will also help determine the size of the compound needed.

> **Site set-up**
>
> 1 Be aware of restrictions on compound areas etc as identified in the planning and/or EIA process.
> 2 Assess the risk of all compound locations and draw up a drainage plan.
> 3 Select locations near existing facilities where possible, for example, foul drainage, water supply, hardstanding.
> 4 Consider a sustainable water supply and minimise water use on site.
> 5 Obtain agreement for wastewater disposal.
> 6 Select locations for cesspits or package plants where no foul connection.
> 7 Store fuel, oil and materials securely and in accordance with legal requirements.
> 8 Select suitable refuelling area(s) on hardstanding with drainage via oil interceptor
> 9 Provide adequate measures to control runoff from compounds and haul routes.
> 10 Provide a suitable vehicle wash area on hardstanding and draining to foul sewer.
> 11 Provide appropriate measures to secure the site and hazardous materials from vandals and trespassers.

16.2 SITE DRAINAGE AND WATER FEATURES

Drainage plans will help in the identification of drains, manholes and outfalls on and off site. If plans are not available, the facilities will have to be assessed. It is essential to know the downstream destination or outfall of both **new** and **existing** drains. Information on upstream sources, features and users is also vital. A drainage plan for the scheme should be prepared that illustrates all surface water bodies and drainage systems within a 500 m-wide corridor of the route, including:

- rivers and streams
- field drains (which tend to flow in wetter conditions)
- foul and surface water drains and outfalls
- ephemeral ditches (which only contain water in wetter conditions)
- canals and other man-made channels
- lakes, ponds, reservoirs and wetlands
- landowners of ditches and waterway banks.

Special attention should be paid to the relationship between natural and artificial networks. Locations where artificial drains discharge to natural watercourses (such as surface water outfall into a river) should be highlighted. The plan will form an integral part of the emergency plan and should be kept up to date (see Chapter 15).

> Use a colour coding system to aid rapid and clear identification of surface water (blue) versus foul drainage (red) networks. Mark on plans and on site with paint, or with painted stakes in areas where drains may become buried.

A linear route may intersect a drainage pipe in several places, so a contractor's activities at one location may affect the drainage on another part of the route where a different contractor may be working. Greater, cumulative effects may in turn be experienced farther downstream.

This is particularly relevant to routes passing through urban or built-up areas, or following existing infrastructure (for example, highway widening). Culverts are often cleansed so that assets may be inspected before work begins, and this may cause a sudden wash of sediments downstream. When working on existing drainage systems, temporary measures should be put in place at the outfall or at the intersection with other drainage to remove sediments and oil. Such measures may include temporarily blocking or diverting the pipe or culvert, or adding a catch pit, sump or geotextile screen.

> **Caerphilly stream pollution costs construction company more than £9000** (source: EA press release, Jan 2005)
>
> A construction company pleaded to a charge of polluting the Nant yr Aber stream, a tributary of the River Rhymney. The company was fined £8000 and was ordered to pay £1400 towards the costs of Environment Agency Wales, which brought the prosecution.
>
> The incident occurred when liquid effluent was released during decommissioning. A biological survey of the affected stream indicated that a 1.1 km length was affected, with the majority of fish being killed. The Agency environment officer said: "The discharge was caused by a misconnection whereby the surface water drain was used rather than the foul sewer for the liquid effluent. This highlights the importance of an accurate site drainage plan to ensure that where discharges are to be made from a site the correct drain is used."

In rural areas, construction works are likely to encounter field drains, which can carry large volumes of water in periods of heavy rainfall. Where the scheme truncates a field drain it is important to divert flows (with permission) to the outfall in a controlled manner to avoid discharging water into the works. See Section 18.7 for information on preventing pollution of existing drainage systems.

Runoff

The potential for pollution from runoff from compound areas, car parks, storage areas and haul routes is high and should be properly considered at site set-up. High-risk receptors such as watercourses and drains will have been identified as part of the site surface water assessment. However, other factors such as rainfall, ground conditions and soil type also need to be taken into account. A sustainable approach to drainage should be considered for managing the drainage from large areas of hardstanding, for example by using permeable surfaces, French drains and swales. For fuel and storage areas the use of an oil interceptor may be appropriate on long-term projects.

Measures to minimise and control the risk of contaminated runoff from compounds and haul roads should be implemented as part of site set-up. Factors to consider include:

- constructing compounds and haul roads using permeable materials (only suitable where groundwater is not sensitive)
- providing gullies or ditches either side of haul roads to intercept and control runoff
- use of SUDS such as rainwater harvesting, retention ponds and other methods
- installing wheel-wash facilities with a dedicated drainage system (see also Section 16.4.4)
- constructing temporary haul road bridges over watercourses to avoid fording.

See Chapter 18 for guidance on controlling runoff and sedimentation.

16.3 WATER SUPPLY

Construction projects need a supply of clean freshwater for use as drinking water, for welfare facilities and for specific construction activities such as concrete batching, wheel-washing etc. It is important to consider the water supply needs of the project and identify potential water supplies early on. Water suppliers, environmental regulators, landowners and local businesses should be contacted as appropriate at an early stage so that the needs of local water users as well as those of the project can be met.

Water supplies may be provided by a direct connection to a mains supply, tanker or abstraction from local borehole or surface water depending on the location of the project. Where a mains water supply does not exist, it should not be assumed that water can be abstracted from surface water or groundwater. A licence is likely to be required for abstractions above 20 m^3/day, but there can be seasonal restrictions, while in the south of the UK abstraction may not be allowed at all because of over-demand. Alternative solutions include:

- transferring ownership of an abstraction licence from someone who is not using it
- purchasing water from someone with an abstraction licence (eg local farm supply).

Whatever the water source, it is likely that a charge will be made for the volume of water used, and ultimately the volume disposed. It is cost-effective to establish the most sustainable supply and to get the set-up correct. Mains water should only be used for drinking water. Mains water has been treated to reach European drinking water standards – water of this quality is wasted when used for dust suppression or washing down construction plant and equipment. In addition, the use of chlorine to treat mains water supply also means it can have a high pollution potential if discharged to surface waters. Settlement ponds, tanks or dewatering operations on site can provide a good source of water for general construction activities such as concrete washout, dust suppression and vehicle washing. Opportunities for water conservation, reuse and recycling should always be considered, as they provide both economic and environmental benefits.

16.4 WATER USE

16.4.1 Site offices and compounds

If the main site offices are going to be in use for several years it will be worth implementing water conservation measures. The key guidance below indicates sources of information providing useful advice and simple measures to reduce water use, and costs, within the construction industry. The DTI has developed key performance indicators (KPIs) to benchmark the use of water (as well as waste, energy and other criteria) on site against that of the construction industry nationally. The use of sustainable drainage schemes (SUDS), such as rainwater harvesting (collecting and using clean runoff from roofs and hardstanding), is encouraged.

> **Key guidance**
>
> Construction Products Association (2005). *Construction products key performance indicators. 2005 handbook* (updated annually):
> <www.constrprod.org.uk>
> <www.dti.gov.uk/construction>
> <www.constructingexcellence.org.uk/resourcecentre/kpizone>
>
> Environment Agency (2001a). *Conserving water in buildings*, series of fact cards, <www.environment-agency.gov.uk/subjects/waterres/>
>
> Envirowise (2005). GG067 *Cost-effective water saving devices and practices*
>
> Martin et al (2000a). C521 *Sustainable urban drainage systems – design manual for Scotland and Northern Ireland*
>
> Martin et al (2000b). C522 *Sustainable urban drainage systems – design manual for England and Wales*
>
> Rainharvesting Systems, advice on products and designs for reusing rainwater, <www.rainharvesting.co.uk>
>
> Scottish Water (2005), "10 Top Tips", 10 guidance leaflets for business, <www.scottishwater.gov.uk>
>
> Water Service in Northern Ireland, <www.waterni.gov.uk>

16.4.2 Concrete washout

A dedicated area should be provided for washing out ready-mix concrete lorries and equipment, either at an on-site batching plant or, if batching is off site, near the works area. Where possible, water from settlement ponds or similar facilities should be used or washout water recycled. Once the solids have settled out, the water can be recovered for use in subsequent batches of concrete (subject to it being chemically suitable), or it may be reused for washing out trucks to minimise the volume requiring disposal.

See Chapter 22 for further essential guidance on concrete and grouting activities.

16.4.3 Dust suppression

Water from settlement ponds can be pumped into a bowser and used to damp down haul roads and site compounds to prevent the generation of dust. Care must be taken to avoid clouds of vapour being created near people, which otherwise could encourage the spread of Weil's disease (leptospirosis) through inhalation. Water pumped from surface or groundwater for use in dust suppression may need an abstraction or water transfer licence (see Chapter 13). Silty or oily water should not be used for dust suppression, because this will transfer the pollutants to the haul roads and generate polluted runoff or more dust. Water bowser movements need to be monitored, as the application of too much water may lead to increased runoff.

16.4.4 Vehicle washing

Wheels are often washed before vehicles leave the site to prevent the build-up of mud on public (and site) roads. Commercially available wheel-washes take water from the mains supply and generally remove silts, oils etc before recycling it and ultimately discharging it via the foul drain. Those with the most efficient water recycling not only conserve water but also cost less to run.

Manned jet washes or lance sprays should only be used in a bunded area where the runoff can be contained and channelled to a treatment area, such as a settlement pond, prior to discharge (see Chapter 13). To prevent the possible spread of Weil's disease through inhalation, care should be taken not to create clouds of vapour from non-potable water supply (eg recycled from settlement pond) near people. Runoff from wheel-washes and vehicle wash bays must not be allowed to enter surface water or foul water drainage systems without permission.

> **Key guidance**
>
> EA, SEPA and EHS (2001). PPG13 *High pressure washers and steam cleaners*
>
> EHS (2000). *Water (Northern Ireland) Order 1999. Guide for vehicle wash operators, treatment and disposal options*
>
> WheelWash, <www.wheelwash.co.uk>

16.4.5 Commissioning

Pressure testing (hydrotesting) and commissioning of pipelines often require large volumes of water. A discharge consent may be required where the water used in commissioning is to be discharged to surface water. It is important to ensure that the water is safe to discharge – water from pipeline testing and, in particular, from commissioning water mains, often contains a high concentration of chlorine used in the sterilisation process. This can have a serious environmental impact when discharged to watercourses.

An abstraction licence may also be required if the water to be used is sourced locally from surface or groundwater. The rate of abstraction as well as the volume may be restricted, and the regulator is likely to require the abstraction to be monitored.

Early discussion and agreement with the environmental regulator is advised, particularly on long pipeline runs. See Chapter 13 for more details on obtaining licences and consents.

> **Water supply and use**
>
> 1   Consult the water company, landowners and environmental agency early on to source a sustainable supply of water for the site.
>
> 2   Obtain the necessary permissions and licences for a water supply and for its disposal.
>
> 3   Conserve water where possible – recycle water from settlement facilities for vehicle washing and dust suppression.
>
> 4   Implement SUDS techniques such as settlement ponds, swales or rainwater harvesting to manage surface runoff where site offices and compounds are established for a period of time.

16.5 WASTEWATER DISPOSAL

Where mains sewerage is available, welfare facilities should be put in an appropriate location and connected to the sewerage system. Agreement from the sewerage undertaker will be required (see Chapter 13).

Where mains sewerage is not available, or where the compound is small and/or temporary, cesspools, septic tanks or package plants can be installed and appropriate discharge agreed with the environmental regulator.

Cesspool – a covered watertight tank used for receiving and storing sewage that has no outlet. A tanker is needed to remove the sewage for onward treatment at a sewage works. This method is considered to be the least sustainable option, as it involves the collection and disposal of wastewater by road tanker. A cesspools may nevertheless be suitable for a short-term temporary arrangement.

Septic tank – a two- or three-chamber system that retains sewage for a sufficient time to allow the solids to form into a sludge at the base of the tank, where it is partially treated. The remaining effluent drains from the tank by means of an outlet pipe direct to ground via a soakaway. Septic tanks need to be regularly desludged and serviced, the interval depending on the size of the tank and the number of personnel it serves. The installation of septic tanks is often not permitted in groundwater protection zones or near wells or boreholes. Guidelines on the use of septic tanks are provided in DETR Circular 03/99 and by CIRIA (see key guidance below).

Package sewage treatment plant – often a self-contained unit used for the full treatment of sewage or a unit that treats wastewater effluent from septic tanks to a higher standard. Package sewage treatment plants are costly to purchase and install and often are best suited to large, long-term projects with large numbers of personnel. The initial cost is often offset by the lower operating cost in the long term. As the effluent is treated to a high standard, water can often be discharged direct to a watercourse or to a soakaway (with permission).

> **Key guidance**
>
> CIRIA (1998a). SP144BT *Septic tank systems: a regulator's guide*
>
> CIRIA (1998b). SP144L2 *Septic tank systems, 2: options*
>
> CIRIA (1998c). SP144L3 *Septic tank systems, 3: design and installation*
>
> Dee and Sivil (2001). PR72 *Selecting package wastewater treatment plants*
>
> DETR (1999). Planning Circular 03/99 *Planning requirements in respect of the use of non-mains sewerage incorporating septic tanks in new development*
>
> EA, SEPA and EHS (2004b). PPG4 *Disposal of sewage where no mains drainage is available*

16.6 STORAGE AND USE OF MATERIALS

16.6.1 Legal requirements

The Control of Substances Hazardous to Health Regulations 2002 (COSHH) and the Management of Health and Safety at Work Regulations 1999 control the use and storage of hazardous substances on site. COSHH assessments must be made for all substances on site and measures identified and implemented to prevent or control exposure to the workforce.

> **COSHH assessments**
>
> Together with the materials safety data sheets (MSDS), these assessments are important with respect to environmental protection as they help determine:
>
> - the substance and its hazardous properties
> - potential risks to the environment
> - storage arrangements
> - disposal arrangements
> - emergency arrangements and what to do in the event of a spill.

The Control of Pesticides Regulations 1986 (as amended) (COPR) governs the storage and use of herbicides and pesticides. Consent is required for use, supply and storage, by the Pesticides Safety Division of Defra.

The Groundwater Regulations 1998 were introduced to prevent the careless use and disposal of substances that could pollute groundwater. The regulations identify two categories of controlled substances known as List I and List II substances. List I substances include mineral oils and hydrocarbons. More information is available at <www.defra.gov.uk/environment/water/ground/guidance.httm>.

16.6.2 Ordering, storing and handling materials

Ordering

The first step in trying to reduce the pollution potential of materials delivered, stored and used on site is to consider whether they are required at all. It may be possible to use alternative materials that are more environmentally considerate and less polluting (see Section 16.8.3 Biodegradable oils). Second, it is worth considering the application of the products and where they will be used and stored. It may be possible to undertake these activities off site or at a less sensitive area along the route.

If possible, only small quantities should be stored on site and materials ordered only when needed. This will cut down the length of time materials have to be stored and reduce the risk of pollution, as well as potential for damage and theft.

A member of the site team or a storeman with responsibility for ordering materials and maintaining stock levels should be appointed on each project, regardless of size. On linear projects there may be a need to appoint a storeman at each compound.

Deliveries

Before starting work on site the delivery arrangements for bulk materials, fuel, oils, chemicals etc should be discussed with the key suppliers and, where appropriate, with subcontractors. At the same discussions, safe access routes and emergency procedures can be agreed, particularly for deliveries taking place at remote locations where the access route may be either off the main road, narrow, across a watercourse, muddy and/or rutted. There will need to be an assessment of the benefits and risks associated with having smaller, more frequent deliveries, against those associated with less frequent but larger deliveries. The storeman, or a member of the site team, should supervise all deliveries to site.

Storage

Arrangements for the storage of materials, containers, stockpiles and waste, however temporary, must be in accordance with legal requirements and where appropriate, follow best practice at all times and at all work sites along the route.

Wherever possible, storage areas should be located:

- well away from sensitive receptors (watercourses, aquifers, drains etc) – at least 50 m from a spring or borehole and 10 m from a watercourse or drain
- on level ground
- on an impermeable base – concrete slab or other areas of hardstanding
- under cover to prevent damage from the elements
- in secure areas
- well away from moving plant, machinery and vehicles.

Containers used to store materials such as fuel, hydraulic oils, chemicals, solvents etc must be fit for purpose and of sufficient strength and structural integrity to ensure that they will not fail or leak. All containers should be stored upright and clearly labelled with capacity and contents (in accordance with COSHH) and appropriate hazard warning signs displayed. Consideration should also be given to additional protection arrangements such as bunded storage areas or the use of drip trays (see Section 16.8) especially where the risk from leaks or spillage is particularly high (eg storage compounds located on groundwater protection zones).

It is important to establish and maintain an up-to-date inventory of the type of product stored/used and the quantity available on site. This is particularly true for hazardous materials that have the potential to cause damage or harm to the environment. The information contained within the inventory is also particularly useful to the emergency services in the event of accidental spillage or other emergency such as fire. The inventory should include the following as a minimum:

- product type (eg solvent)
- trade name
- UN number
- maximum quantity stored
- location of material on site
- material safety data sheet (MSDS) or COSHH assessment

Areas designated for the storage of stockpiles should be on land reinstated in the early phases of a project or in areas not required until later in the project. Double-handling is not only an added cost, but it also raises the risk of pollution incidents occurring by increasing vehicle movements and exposing soil to rainfall, and necessitates the use of additional sealing or bunding materials. See Chapter 18 for advice on preventing runoff from stockpiles.

Ordering, handling and storing materials

1. Can the use of potentially polluting materials be eliminated from the construction process?
2. Can processes be undertaken elsewhere at less sensitive locations, for example off site?
3. Can alternative materials that are more environmentally considerate be used, for example, biodegradable oils?
4. Check stock requirements – can smaller quantities be stored on site or delivered to site when they are needed?
5. Identify how materials will arrive so that the appropriate arrangement for handling and storage can be made.
6. Ensure deliveries are supervised at all times.
7. Store materials in accordance with manufacturer's requirements.
8. Store materials away from drains and watercourses.
9. Store materials away from vehicle movements.
10. Consider the effects of extreme weather conditions for material storage.
11. Provide appropriate containment for hazardous materials.
12. Maintain an up-to-date inventory of products/materials stored.
13. Provide waste storage areas at all work sites where possible.

16.7 WASTE MANAGEMENT

16.7.1 Legal requirements

The Environmental Protection Act 1990 introduces the duty of care on all those who produce, keep, treat or dispose of waste. The Environmental Protection (Duty of Care) Regulations 1991 (as amended) in England, Wales and Scotland and the Controlled Waste (Duty of Care) Regulations (Northern Ireland) 2002 require that all wastes are stored correctly, are handled only by businesses that are licensed to do so and that any wastes transferred are accompanied by written documentation. A code of practice for the duty of care is available from Defra and EHS.

Because of their harmful properties, hazardous wastes – including waste oils, chemicals, solvents – are subject to more stringent waste management legislation, including the Special Waste Regulations (Northern Ireland) 1998, the Special Waste Amendment (Scotland) Regulations 2004 and, most recently, the Hazardous Waste (England and Wales) Regulations 2005. Under the latter Regulations, sites in England and Wales producing hazardous waste – eg waste oil, empty oil, chemical and paint containers, contaminated land – must register with the Environment Agency.

In July 2004 the DTI published *Site waste management plans. Guidance for construction contractors and clients*. This is intended as a voluntary (at time of writing) code of practice aimed at medium-sized to large companies involved in any building or civil engineering project with a total value of £200 000 or more. As a result, most government clients (eg Highways Agency, Environment Agency, Ministry of Defence and local authorities) require site waste management plans as part of contract requirements.

16.7.2 Waste handling and storage

Where appropriate, waste should be stored and handled in accordance with the site waste management plan (see above).

Sufficient waste storage should be supplied near to all working areas, as well as to the main compound, so that staff are encouraged to dispose of waste correctly. Where this is not possible, designated storage areas can be established along the route of the scheme. This avoids materials being left around the site and also assists with the collection, reuse and disposal of waste materials.

Wastes should be stored in designated areas that are isolated from surface water drains or open waters. Skips need to be closed or covered to prevent materials being blown or washed away and to reduce the likelihood of contaminated water leakage. Hazardous wastes, such as waste oil, chemicals and preservatives, must be stored in sealed containers and kept separate from other waste materials while awaiting collection by a registered waste carrier. Recent requirements under the Hazardous Waste Regulations 2005 also make it an offence in England and Wales to mix hazardous and non-hazardous wastes, and to mix hazardous wastes of different categories (eg mineral oil with oily water or petrol with engine oil). For details, see the List of Wastes (England) Regulations 2005.

> **Key guidance**
>
> Defra (2005). *Guidance on mixing hazardous waste*
>
> DoE (1996). *Waste management – the duty of care: a code of practice*
>
> DTI (2004). *Site waste management plans – guidance for construction contractors and clients*, <www.dti.gov.uk/construction/sustain/site_waste_management.pdf>

16.8 FUEL AND OIL

16.8.1 Storage and legal requirements

Oil and fuel are the most frequently recorded types of water pollutant investigated by the Environment Agency in England and Wales. The Control of Pollution (Oil Storage) (England) Regulations 2001 were introduced to control the storage of fuels and oil (including diesel, hydraulic oils, mould oils, mineral oils, petrol and heating oils) and to help cut the number of oil-related pollution incidents. The regulations are applicable only to the storage of oil in England, although similar legislation is being considered in Wales, Scotland and Northern Ireland. Contractors working in these areas of the UK are advised to comply with the best practice requirements identified below.

The regulations are applicable to any oil container with a storage capacity of 200 litres or more. A 45 gallon drum = 205 litres.

In summary, the regulations require:

- containers to be stored within a secondary containment system (eg a bund for static tanks or a drip tray for mobile stores and drums)
- bunds must be capable of storing 110 per cent of the tank capacity. Where more than one tank is stored, the bund must be capable of holding 110 per cent of the largest tank or 25 per cent of the aggregate capacity (whichever is the greater)
- ancillary equipment, such as vent pipes, delivery pipes, sight gauges, valves and refuelling hoses, must be contained within the bund
- taps, nozzles or valves through which the oil is dispensed must be fitted with a lock and locked shut when not in use
- drip trays used for drum storage must be capable of holding at least 25 per cent of the drum capacity. Where more than one drum is stored the drip tray must be capable of holding 25 per cent of the aggregate capacity of the drums stored.

Key guidance

EA, SEPA and EHS (2004c). PPG2 *Above ground oil storage tanks*

EA, SEPA and EHS (2004d). PPG7 *Refuelling facilities*

EA, SEPA and EHS (2004e). PPG8 *Storage and disposal of used oils*

EA, SEPA and EHS (2004f). PPG26 *Storage and handling of drums and intermediate bulk containers*

EA, SEPA and EHS (2004g). *Getting your site right – industrial and commercial pollution prevention*

EA et al (2000a). *Masonry bunds for oil storage tanks*

EA et al (2000b). *Concrete bunds for oil storage tanks*

Mason et al (1997). R163 *Construction of bunds for oil storage tanks*

Teekaram et al (2002). C535 *Above-ground proprietary prefabricated oil storage tank systems*

Fuel tanker – on large earthworks schemes use of a four-wheel-drive, road-going, double-skinned fuel tanker should be considered, as this may eliminate the need for several storage tanks and bowsers and may contribute to security of fuel storage. Access to sites along linear schemes will often necessitate the use of public roads. Drivers must have an ADR certificate to transport fuel on public roads. The ADR driver training regulations apply to vehicles over 3.5 t carrying hazardous goods above certain load limits. Below these limits drivers must still receive specific awareness training.

Mobile bowser – a towable storage tank, fitted with fuel-dispensing equipment (Figure 16.1). Frequently used on linear projects to refuel plant remote from the site compounds, bowsers themselves are refuelled at the main fuel stores. Modern bowers are often integrally bunded, as described below. The use of unbunded (or "single-skinned") bowsers is illegal in England.

Figure 16.1 *Mobile bunded bowser*

Common sense should prevail in the selection of plant to tow a bowser and strict controls on vehicle speed limits for bowsers and tankers need to be in place. Rutted, muddy ground should be avoided and site access routes or haul routes improved where necessary to avoid damaging or overturning the tanker or bowser (see Transco case study below). Particular care is needed where the haul route crosses a watercourse – a temporary bridge or culverted crossing is preferable to regular fording of a watercourse. Where bowsers travel on public roads, additional regulations under the transport of dangerous goods may apply (see "Fuel tanker" paragraph, above). Mobile bowsers should be returned to a secure, impermeable storage area, away from drains and open water when not in use.

Case study – red diesel spillage (source: Transco)

On a pipeline installation site a refuelling tanker ran over some site debris, which fractured the underside fuel off-take line, resulting in the fast, uncontrolled release of 1000 gallons of red diesel. Site staff responded well to the emergency and used a machine to create an earth bund, which allowed the recovery of some 500 gallons of free product. The area was cleaned by the disposal of the contaminated soil to an appropriately licensed landfill.

Learning points:

- accidents are preventable
- site staff should be well trained on emergency procedures
- assess the susceptibility of fuel vehicles to pipework damage
- maintain good housekeeping in the working area
- prohibit refuelling in areas such as source protection zones.

1000 gallons of red diesel contained with earth bund prior to pumping out

Integrally bunded tank – a tank located within a bund that has roof and sides built in (Figure 16.2). The bund forms an integral part of the tank, enclosing all the ancillary equipment such as valves and hoses. Integrally bunded tanks often come supplied with hand pump for dispensing fuel, but larger tanks may have a small electric or petrol pump. This type of tank prevents rainwater ingress and has lower maintenance requirements as a result. It also provides more secure fuel storage on remote areas of the site.

Figure 16.2 *Integrally bunded tank (courtesy Nuttall)*

Open bunded tank – a tank, often gravity-fed, that sits within an open bund usually constructed of masonry blockwork or concrete (Figure 16.3). The bund walls and base must be well constructed, of sufficient strength and be impermeable to water and oil – either rendered or coated with a sealant that will not be damaged by the contents stored. There should be no holes or outlets in the bund wall. A roof or cover will prevent collection of rainwater, which can compromise the storage capacity of the bund and be time-consuming and expensive to pump out and dispose of.

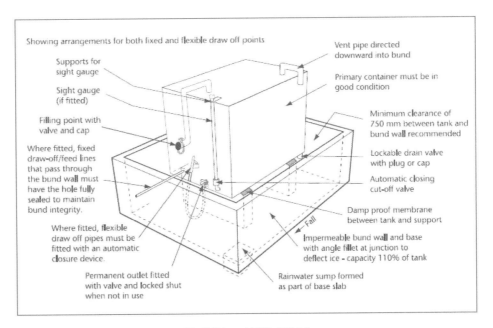

Figure 16.3 *Open bunded tank (EA, SEPA and EHS, 2004c)*

Drums – must be stored upright. Oil drums with a storage capacity of 200 litres or more must be stored within a drip tray (or sump pallet). However, any oil container (regardless of size) should be stored in a drip tray if the contents pose a high risk to the environment. Designated drum storage areas should be established at defined locations about the site (see Figure 16.4). Ideally, these need to be covered and well secured to avoid the risk of damage or vandalism. Guidance in the environmental regulators' PPG26 *Storage and handling of drums and intermediate bulk containers* (EA, SEPA and EHS, 2004f) should be followed.

Figure 16.4 *Storage site guidelines (courtesy Nuttall)*

Inspection and maintenance

Fuel and oil stores, including tanks and drums, should be inspected regularly for signs of spillages, leaks and damage during use. A record should be kept of these inspections and any improvements needed, which should be carried out immediately.

Rainwater and debris are likely to accumulate in open bunds and drip trays, reducing their capacity. The volume of rainwater should not exceed 5 per cent of the total capacity of a bund or drip tray. Rainwater can be discharged to the foul sewer or to ground provided no oily sheen is visible. Small quantities of oil can be removed from the surface of the water using oil-absorbent pads or water poured from drip trays etc through a pad before disposal. Where the oil/water mix cannot be separated fully the liquid needs to be removed and stored before disposal by a specialist waste management contractor.

16.8.2 Handling and refuelling

The risk of spilling fuel is at its greatest when refuelling plant. Only designated trained and competent operatives should be authorised to refuel plant on site. Plant and equipment should be brought to a designated refuelling area rather than refuelling at numerous locations about the site, particularly if they are in high-risk areas. Refuelling areas should ideally be on hardstanding with provision for the prevention of fuel discharging off site. This often involves the use of an adequate oil separator with appropriate discharge consent. The hardstanding should be large enough to catch spillages, but as small as possible to minimise the volume of rainwater that the separator has to deal with. Where this is not possible additional arrangements can be made, such as the use of drip trays, mats etc, when refuelling.

Operatives need to be made aware of the potential risks associated with refuelling, oil changes, hydraulic oil and the use of other oil-based products (mould oil) and told what to do in the event of spillage.

> **Case study – misconnection causes fuel leak** (source: Nuttall)
>
> Approximately 1000 litres of diesel fuel were lost to ground on a major civil engineering project as a result of poor labelling and poor location of a static fuel tank. A tanker delivery driver connected the delivery hose to the vent pipe (which was not labelled) instead of the filler pipe. The fuel tank was poorly located and could not be seen during refuelling. The delivery of the fuel caused a build up of pressure within the tank leading to catastrophic failure at the base weld of the tank. The hydrostatic pressure within the tank caused the fuel to jet over the surrounding bund. The resulting clean-up costs were significant.

Emergency arrangements

Procedures and contingency plans ought to be in place at each work site to address cleaning up small spillages as well as dealing with an emergency incident. A stock of absorbent materials such as sand, spill granules, absorbent pads and booms should be kept at each work site, on plant working near water and particularly at refuelling areas and where fuel or oil is stored. A spill response procedure should be set up and staff trained to deal with spillages including the use of spill kits. See Chapter 15 for further guidance on emergency and contingency planning.

16.8.3 Biodegradable oils

In high-risk areas, particularly for work taking place on or near a watercourse, environmentally considerate lubricants, such as synthetic non-toxic biodegradable hydraulic fluids, should be used. Since June 2005 the Environment Agency for England and Wales requires its contractors to use only biodegradable hydraulic fluid in tracked excavators.

Table 16.1 *Advantages and disadvantages of biodegradable oils*

Advantages	Disadvantages
• Environmentally considerate; readily biodegradable and non-toxic to flora and fauna • extended service and maintenance intervals; fewer oil changes; cost savings by avoiding need for regular purchase and replacement of old mineral oil • miscible with existing mineral-based oils, so can be retrofitted into existing hydraulic systems without the need for modification • reduced risk of pollution of the aquatic environment; complies with all government policies, Defra and CIRIA guidelines on sustainable development • can be used in a wide range of mobile and fixed plant on construction sites	• More expensive to purchase than a mineral-based oil • hydraulic system must be well maintained to prevent leakages or ingress of dust, dirt and water • oil must be condition-monitored regularly • environmental benefits are reduced if biodegradable oil is cross-contaminated with more than 5 per cent mineral oil

> Biodegradable oils must be stored, handled and disposed of as any other oil and must not be disposed to surface water or foul drains.

> **Fuel and oils**
>
> 1. Assess the risks of potential oil storage areas.
> 2. Establish fuel and oil stores in line with the Control of Pollution (Oil Storage) (England) Regulations 2001.
> 3. Train operatives in refuelling plant, equipment and the application of oil-based products.
> 4. Develop emergency procedures for dealing with spills and communicate these to the workforce.
> 5. Ensure spill materials are readily available on site.
> 6. Regularly inspect and maintain oil stores, bunds and drip trays.

16.9 SITE SECURITY

Preventing unauthorised access and acts of vandalism is essential on all construction projects and careful consideration is an integral part of the site planning and set-up process. Acts of vandalism and poor site security have been cited in a number of water pollution cases on major construction projects. Intruders on site often cause damage that can result in water pollution incidents by:

- opening taps and valves on fuel tanks
- tipping over drums and other containers
- stealing or moving raw materials, plant and equipment
- destroying temporary works and work in progress.

> **Contractors can be liable for environmental damage caused by vandals if they have not made reasonable attempts to guard against it. This liability may increase if vandals have struck at a site on a previous occasion.**

On linear projects, securing the whole site can be difficult because the length of the route, multiple work locations, numerous access points and remote satellite compounds. Early assessment of the sensitivity of a project and identifying potential locations at risk will assist in the design of the site layout and security measures needed.

> **Vandalism leads to £13 000 fine for oil spill** (source: *The ENDS Report*, Aug 2005)
>
> A diesel generator hire company was fined £13 000 for an oil spill that occurred after vandals broke into its premises.
>
> The incident occurred when vandals released 2000 litres of oil from containers on site. A company employee partially cleaned up the spill but failed to prevent oil entering the site drains leading to the Spittle Brook.
>
> Although the company told the Environment Agency that it was aware of its responsibility to store oil safely, it had stored oil close to surface water drains and without any form of barrier to contain a spillage. The site had been broken into three times in the past, but the company had not taken adequate steps to protect against vandalism.

Fencing

Fencing serves two purposes: it defines the site boundary/site ownership and it acts as a physical barrier to deter intruders. Various types of fencing are available. Solid barriers such as hoarding are considered more difficult to scale than chainlink fencing and also prevent casual surveillance of the site by prospective thieves. Hoarding, however, can also provide cover for thieves and vandals once on site.

On linear projects, particularly large-scale open projects such as road schemes, it is not practical to put hoarding along the whole site. Instead post-and-rail fencing (which can be part of the permanent works) may be installed to denote the site boundary, while more secure fencing can be deployed around the site office and compound areas to protect the materials, plant and equipment stored.

Secure access

Strict controls on access and the movement of people on site can help reduce incidents of trespass and theft. Measures that should be considered include:

- the use of site passes or swipe cards
- signing in and out
- accompanying site visitors
- delivery drivers reporting to the main office on arrival
- supervision of all deliveries
- personnel at specific work locations such as footpath crossing
- 24-hour security guards
- installation of CCTV.

Securing plant and equipment

Plant and equipment can be protected from unauthorised use with the following measures:

- locking devices (eg steering lock, clutch pedal lock, mortice security deadlock on cab doors and wheel clamps)
- mesh grilles on cab windows and doors
- parking up plant close together at the end of shifts
- returning heavy plant to the main compound at weekends
- storing high-risk equipment out of sight in lockable containers.

Secure storage of materials

The security of fuel, chemicals and other material storage areas also needs to be considered, both to avoid loss from theft and to reduce the potential for uncontrolled releases from acts of vandalism.

Materials should not be stored against site boundaries and fencing, as thieves and vandals can use them to enter the site. All potentially hazardous materials should be made secure and stored in accordance with COSHH requirements (see Section 16.6), preferably in designated secure areas.

Physical barriers within storage compounds and particularly around fuel tanks can be an effective form of protection against collision from heavy plant and vehicles and provide an additional line of defence to vandals on foot. Barriers should be:

- designed to withstand impacts from the heavy plant on site
- placed a suitable distance from the storage facility to provide a buffer zone
- located where damage is most likely to occur – often at the corners of fuel bunds, areas of restricted space and the entrance to the site compound.

Signage

Warning notices can be displayed around the site perimeter and on the approach to site compound areas to deter potential intruders. Signs should be clear, visible and unambiguously worded.

Security lighting

Site security lighting is a good deterrent to vandals and thieves, but it can also cause nuisance to local residents. Site lighting needs to be kept at the minimum brightness to provide adequate security and safety and angled so that it does not shine directly into residential properties. Consideration should be given to the use of automatic lighting activated at darkness or by infrared sensors that respond to body heat and movement.

Emergency incidents

Emergency arrangements need to take into account the potential for water pollution from acts of vandalism, actions in the event of an incident and contact details outside normal working hours. All site personnel, including security staff, need to be briefed on these arrangements. See Chapter 15 for further details on emergency planning.

> **Case study– Security guard raise alarm after diesel tank fails at plant depot** (source: Nuttall)
>
> During the night a delivery hose failed on a static tank providing oil for workshop heaters in a plant maintenance depot. The oil was contained within the bund. Audible alarms were activated when the oil in the bund reached a critical level. On hearing the alarm the site security guard was able to isolate power to the diesel tank and deal with a small spill of oil that had overtopped the bund, as set out in the site emergency arrangements.

17 Adjacent land and water use

A linear site is likely to occupy a narrow strip that intersects land in many different uses and will have a large number of "neighbours". The interaction between the site and the adjacent land uses is critical because, while the site cuts a neat line across the land, surface water and groundwater will not recognise this boundary and will continue to flow both into and away from the site works. Sources of water and pollution off site are just as likely to affect the site as the site itself is to introduce polluted water into the surrounding environment. In addition, linear sites are typically restricted in terms of land take, so adjacent land may provide suitable areas for water containment and treatment in the event of an emergency.

17.1 PROTECTING ADJACENT LAND AND WATER USES

Where available, the environmental statement (ES) prepared for the scheme should be consulted to identify all the sensitive environmental receptors and key users of the land and water assets surrounding the site (Table 17.1). It is particularly important to consider those downstream. For construction projects where a specific ES has not been prepared, or the information in the ES does not clearly identify sensitive receptors outside the site, details can be obtained from the environmental regulator, local conservation agencies, landowners, local site staff knowledge and personal observation.

Table 17.1 *Sensitive receptors*

Surface water	Groundwater
Fisheries	Abstraction points – private borehole, farm supply, water company, industrial
Water abstraction – farm, water company, industrial	
	Aquifer
Recreational river – sailing, canoeing, fishing etc	Source protection zone
Visible to public – residential area, park etc	Ponds and other surface waters affected by groundwater level
Protected site – SSSI etc	Wetlands
Protected species – lamprey, eel	

> Sensitive receptors may be some distance downstream of the site.

Where key concerns and sensitive sites have been identified it is important for a member of the project team to make contact to understand stakeholder needs and to ensure that appropriate mitigation measures are in place. Landowners and occupiers should be advised about what is being done to protect their interests. If necessary, their contact details can be included in the emergency plan, so they can be informed as soon as possible should there be a pollution incident. Landowners and occupiers also need to know whom to contact on site if they have any queries or complaints and should be given a prompt response.

Case study – railway works cause pollution of downstream fishery

A civil engineering contractor pleaded guilty to the charge of allowing polluting matter (namely silt) to enter a controlled watercourse contrary to Section 85 (1) of the Water Resources Act 1991. The company was fined £10 000 plus £2134 costs.

Figure 17.1   Silt pollution at fish farm

The Environment Agency (EA) was alerted to the incident after receiving complaints from a trout farmer located about 1 km downstream of the construction works who reported a high level of silt in the water passing through his nursery (Figure 17.1).

The investigating EA officer traced the plume of silt upstream (Figure 17.2) to a site where the contractor was installing new trackside drainage in a railway cutting.

On arrival he found that the pollution prevention measures in place had been overwhelmed following a prolonged period of heavy rain (Figure 17.3).

Figure 17.2   Silt pollution in river

Construction activity had already been suspended and emergency arrangements implemented in order to contain and remove the silt using tankers. The EA officer recognised the clean-up efforts of the site team (Figure 17.4), but concerns were raised that neither the EA nor downstream water users (in particular the fish farm affected) had been notified of the problem.

Further investigation into the incident by the contractor identified a number of contributing factors. The incident showed there was a need to:

- understand the geology and hydrology of a site when planning the work
- consider the potential impact of water pollution on stakeholders downstream of the works
- monitor weather conditions closely
- review risk assessments, method statements and pollution prevention measures when the method of work or conditions on site change
- liaise with the EA and key stakeholders early in the contract and maintain this throughout the project
- notify the EA and other stakeholders downstream immediately should there be a pollution incident.

Figure 17.3   Construction works

Figure 17.4   Improved working area

Pollution prevention measures should be in place to protect the interests of adjacent users. Downstream users could suffer cumulative impacts of numerous discharges or water crossings on the route. In the event of an incident, the cost of cleaning up a sensitive site can far exceed the fine. Such costs may include restocking a fishery or cleaning up a contaminated aquifer and can be up to 20 times the cost of installing prevention measures. Pollution prevention measures are discussed throughout this guidance – see Table 17.2.

Table 17.2 *Pollution prevention measures – summary and further guidance*

Key issues in protecting adjacent users	Refer to chapter/section
Do not discharge to surface waters or drains without obtaining consent from the environmental regulator and landowner. Ensure discharge water quality is treated to meet the required standards; strict water quality standards may be imposed if the watercourse is a protected site, a fishery or used for abstraction.	Surface water, 2.1 Water treatment and disposal, 19 Licence and consents, 13
Control and treat surface water runoff before leaving the site. Ensure on-site operations such as constructing earth bunds, pumping or drainage works do not cause or worsen flooding on neighbouring land.	Runoff and sediment control, 18 Water treatment and disposal, 19
Select suitable methods of work for temporary and permanent works in or near watercourses. Permission may be required from the environmental regulator.	Works in or near water, 20 Licence and consents, 13
Prevent dust and litter being blown by the wind.	Site set-up, 16
Only dewater groundwater volumes in accordance with a licence from the environmental regulator.	Licences and consents, 13 Groundwater, 2.2 Excavations and dewatering, 21
Have a contingency plan in the event of an emergency.	Emergency and contingency planning, 15
Inspect and/or monitor accessible off-site areas regularly so that problems can be identified quickly. Keep a record of inspections.	Monitoring, 14

17.2 PROTECTING THE SITE FROM ADJACENT ACTIVITIES

All the potential sources of pollution around the site, particularly those upstream, should be identified from either the ES or discussion with the environmental regulator. When off-site runoff is being diverted away from the site (Chapter 18) or to a suitable outfall, it is important to clarify who is responsible for any off-site pollution such as agricultural runoff that may be released through the site's outfall. Records should be kept of activities taking place around the site, even those some distance away such as other construction work along the same watercourse or drainage network, or farming activities up-slope of the site. These records may help in the event of a complaint or an incident.

Measures will need to be put in place to protect the works from off-site pollution and consideration given to the effects of heavy rainfall, flooding or a pollution incident on an adjacent site. See Table 17.3 for more information. Pollution prevention measures are discussed throughout this guidance.

Table 17.3 *Protection from off-site pollution – summary and further guidance*

Key issues in protecting the site	Refer to chapter/section
Be aware of other activities up-slope or upstream of the site, or any contaminated land adjacent to the site. Isolate the site from off-site runoff by installing cut-off drains or ditches to channel water around the site.	Runoff and sediment control, 18 Contaminated land, 23
Divert watercourses around the site or culvert beneath the site where possible and with required permissions.	Surface water, 2.1
Only dewater groundwater volumes in accordance with a licence from the environmental regulator. Over-abstracting water may draw in contaminated water from outside the site, leading to treatment and disposal problems.	Licence and consents, 13 Groundwater, 2.2 Excavations and dewatering, 21
Have a contingency plan in the event of an emergency.	Emergency and contingency planning, 15

17.3 ADDITIONAL LAND TAKE

Keeping contact details of site neighbours will help if there is a need to discuss the possibility of using their land in an emergency. At rural sites, it may be possible to discharge small volumes of silty water on to adjacent land to allow it to infiltrate through the ground (see Chapter 19.2.1). The land may also be used to construct an emergency retention pond if heavy rainfall exceeds the capacity of existing water treatment facilities, or if dewatering volumes are greater than those expected. See Chapter 19 for guidance on methods to treat and dispose of water.

> **Checklist for action – adjacent land use**
> 1. At an early stage, identify sensitive receptors, land and water users from the ES, environment and conservation agencies, local information and observation.
> 2. Notify interested parties, understand their needs and confirm control and emergency arrangements with them.
> 3. Identify off-site sources of pollution and isolate the site where possible.
> 4. Identify areas of adjacent land that may be suitable for water containment or treatment in an emergency.
> 5. Implement pollution prevention measures.
> 6. Monitor site and adjacent activities – see Chapter 14 for details on monitoring.

18 Runoff and sediment control

18.1 INTRODUCTION

Sediment, including all soils, mud, clay, silt, sand etc, is the single main pollutant generated at construction sites and largely arises from the erosion of exposed soils by surface water runoff. The adoption of appropriate erosion and sediment controls during construction is essential to prevent sediment pollution. This chapter is concerned with reducing and managing the quantity of runoff and sediment on site through:

- sediment and erosion control plans
- estimating runoff
- planning for flood conditions
- estimating sediment generation
- erosion and sediment control measures.

It is important to distinguish between erosion control and sediment control:

- erosion control is intended to prevent runoff flowing across exposed ground and becoming polluted with sediments
- sediment control is designed to slow runoff to allow any suspended solids to settle out *in situ*.

A combination of erosion control and sediment control may be required. Once runoff becomes contaminated with sediment it is difficult and expensive to remove. See Chapter 19 for additional methods of water treatment for sediment removal.

> **Principles of erosion and sediment control**
>
> 1. Erosion control (preventing runoff) is much more effective than sediment control in preventing water pollution. Erosion control is less subject to failure from high rainfall, requires less maintenance and is also less costly.
> 2. Plan erosion and sediment controls early in the project and incorporate into the works programme.
> 3. Install drainage and runoff controls BEFORE starting site clearance and earthworks
> 4. Minimise the area of exposed ground.
> 5. Prevent runoff entering the site from adjacent ground, as this creates additional polluted water.
> 6. Provide appropriate control and containment measures on site.
> 7. Monitor and maintain erosion and sediment controls throughout the project.
> 8. Establish vegetation as soon as practical on all areas where soil has been exposed.

Internationally, the requirement for erosion control on construction sites has been part of the legal framework for some years, so there is a body of established guidance from the international construction industry and local authorities.

> **Key guidance**
>
> **International guidance**
>
> Australian Capital Territory (1998). *Erosion and sediment control during land development*, <www.environment.act.gov.au>
>
> Beckstrand et al (2004). *Draft New York standards for erosion and sediment control*, <www.dec.state.ny.us/website/dow/toolbox/escstandards>
>
> City of Houston et al (2001). *Storm water management handbook for construction activities*, <www.cleanwaterclearchoice.org>
>
> International Erosion Control Association, <www.ieca.org>
>
> McLaughlin (2002). *Measures to control erosion and turbidity in construction site runoff*, <www.ncdot.org>
>
> Oregon Department of Environmental Quality (2004). *Best management practices for storm water discharges associated with construction activities*, <www.deq.state.or.us/WQConstructionBMPs.pdf>

18.2 PREPARING AN EROSION AND SEDIMENT CONTROL PLAN

It is essential to plan runoff and sediment controls well before starting work on site. Clients are increasingly requiring contractors to prepare construction site erosion and sediment control plans or similar pollution prevention plans before they begin work.

> **Benefits of planning runoff and sediment controls** (after Hill, 2005)
>
> - Minimises risk of pollution and costly fines
> - reduces risk of delays to construction works from poor or wet ground conditions
> - reduced downtime costs to the contractor, or project promoter if "act of God" is claimed
> - space may be saved and land acquisition costs may be reduced by reducing reliance on settlement ponds
> - quicker, cheaper and better land reinstatement following completion of the works
> - reduces disputes and compensation to landowners who will be able to return land to normal use earlier than heavily damaged land.

The first phase in planning runoff and sediment controls is to collect site information, including details of any protected or sensitive sites identified in the ES or other documentation (see Chapter 12). A site map, preferably topographic, will help in identifying and illustrating existing land use and surface water features on and adjacent to the site (see Chapters 2 and 17). It is important to be aware of the anticipated variation in flow along drainage ways (dry ditches can become torrents after rainfall).

The proposed construction activities should be indicated on the map, including graded slopes, fill areas, stockpiles and locations for storage etc selected using the guidance in Chapter 16.

Following the guidance in Section 18.3, the potential for runoff, ponding of water and even flooding should be assessed by reviewing soil types, topography and rainfall data to ensure that runoff patterns are known before work starts.

Several factors will influence the selection of erosion and sediment controls but site conditions are the most significant. Selection of controls is based on information collected in the site evaluation and runoff estimation stages. Section 18.6 describes in detail the more commonly used controls.

Controls on these additional sources of polluted water (discussed elsewhere in this book) should also be addressed during development of an erosion and sediment control plan:

- vehicle washing
- works in or near water
- storage of construction equipment and materials
- waste management
- water use and disposal.

Once selected, measures to minimise erosion and sediment should be installed and maintained with care if they are to achieve maximum, or indeed any, effectiveness.

> Erosion and sediment controls need to work well until the disturbed soils are stabilised (through vegetation or another method), not just until the end of construction (see the case study below).

Control measures need to be inspected regularly, particularly after rainfall. It is recommended that an inspection and maintenance checklist of the control measures be developed and records kept of the inspections and maintenance activities, such as removal of sediment accumulated at silt traps when they are half full.

Case study – pollution caused from runoff after demobilisation

A civil engineering contractor was found guilty of three offences for allowing polluting matter (silt) to enter a controlled watercourse contrary to Section 85(1) of the Water Resources Act 1991. The company had to pay a total of £40 500 in fines and costs.

The incident occurred towards the end of the construction phase on a large earthworks project. Temporary settlement lagoons and other surface water management measures were removed to allow completion of the permanent works in accordance with the client and design requirements.

A prolonged period of heavy rain caused surface water runoff to erode the newly constructed, unprotected embankments, resulting in silty water entering a river adjacent to the site via the permanent works drainage without any further treatment.

The magistrates found the contractor to be culpable because, as an experienced contractor, it should have foreseen the problems that might arise when the pollution prevention measures were removed.

Developing a runoff and sediment control plan

1. Identify existing land use, surface water features, low-lying areas and natural drainage ways (which may be dry) on and off site and illustrate on a site plan.
2. Identify and illustrate construction activities on the plan.
3. Identify areas most likely to have the potential for runoff, such as steep slopes, cuttings, embankments, stockpiles, haul roads etc.
4. Collect information on soil types and rainfall data and then estimate runoff.
5. Select the best controls to reduce runoff and erosion for the site conditions and control other sources of water on site.
6. Ensure that control measures are correctly installed and adequately sized. Initial runoff controls must be in place before site works begin.
7. Programme inspection and maintenance throughout construction to determine if the controls are working adequately or whether further measures are required.
8. Develop contingency plans for unexpected rainfall and/or flooding; develop contact arrangements with downstream users.

18.3 ESTIMATING RUNOFF

Runoff volume depends on three main factors:

- catchment size and topography
- infiltration rates of the ground
- rainfall intensity and duration.

Estimations of runoff should be undertaken for two purposes.

1. To estimate the volume of water that may require containment in a settlement pond or similar.

2. To estimate the flood flow in streams, ditches and larger watercourses following a rainfall event in order to design temporary works, haul road crossings, culverts etc.

> **Key guidance**
>
> For runoff generated on most linear sites, estimate the volume of runoff using the method for small catchments below (Section 18.3.1), or simply assume all the rainfall (see Table 18.2) will form runoff.

Catchments

There are various ways to estimate runoff. The most appropriate method to use depends partly on the size and type of catchment under consideration and partly on the accuracy required. A catchment is an area of land draining to a certain point such as a river. A linear construction site is likely to cross several catchments of varying size, type and topography, and any runoff will drain across the site. Catchment size is therefore not just the area of the site itself, but includes the surrounding area from which runoff is likely to flow on to the site. Estimations of runoff will have to account for off-site runoff as well as on-site runoff. More information can be found in HA 106/04 "Drainage of runoff from natural catchments" in Section 4.2.1 of the *DMRB* (Highways Agency, 2004).

It can be difficult to estimate the size of catchments, particularly if the topography is flat, urbanised and/or subject to significant man-made alteration. The topography will influence the velocity of runoff and the potential for flooding in low-lying areas. Normally it should be possible to estimate the extent of a catchment from a topographic map, such as an Ordnance Survey map of appropriate scale, but it may be necessary to consult local drainage engineers or a hydrologist.

The larger catchment calculations (Section 18.3.3) indicate the volume of water that can be anticipated to reach watercourses within the catchment and are more appropriate for works in water or floodplains. In these situations, careful planning and consultation with the environmental regulator is essential.

Infiltration rates

The rate of water infiltration through different ground conditions varies. Table 18.1 shows indicative infiltration values. Where the intensity of a rainfall event is greater than the infiltration rate the excess water becomes runoff. This "runoff potential" is illustrated in more detail in Table 18.3. In winter, the ground becomes waterlogged such that there is no infiltration, and all rainfall becomes runoff.

Table 18.1 *Typical infiltration rates for various soils*

Soil texture		Sand	Sandy loam	Loam	Clay loam	Silty clay	Clay
Typical	mm/h	50	25	13	8	3	5
	(range)	(25–250)	(13–75)	(8-20)	(3–15)	(0.3–5)	(1–10)
Typical	l/min/ha*	8000	4000	2000	1300	500	800

* Assuming the water to be spread evenly over the surface (Chant-Hall *et al*, 2005a).

Rainfall

The quantity of rainfall that can be expected at any site in the UK can be obtained from the following sources.

1   Site-specific data obtained for the project (unless the permanent works design requires this data, it is unlikely that it will be recorded).

2   Data for a specific grid reference can be obtained from the CD database that accompanies the *Flood estimation handbook* (IH, 1999).

3   Estimated from mapped information provided by the *Flood estimation handbook* (IH, 1999) and *The Wallingford Procedure* (HR Wallingford and IH, 1981). These reports allow rainfall to be estimated for any duration (eg over one hour, or two days) and for any probability (such as twice a year, or once in 10 years).

4   Estimated from Met Office mapped data of annual rainfall (Figure 18.1); also available at <www.met-office.gov.uk>.

Rainfall events vary with season – heavier short-period rainfall, such as that caused by thunderstorms, can usually be expected in summer, while higher rainfall totals over periods of a day or more are more likely in winter. The most appropriate totals, duration and probability of rainfall will depend on the duration of the works, the severity of the consequences of exceedance and the level of acceptable risk.

For short-term construction works, seasonal estimates of rainfall *totals* should be made, but for the majority of major linear schemes (lasting a year or more), the *annual total* rainfall illustrated in Figure 18.1 should be used.

It is suggested that a rainfall *probability* of at least once in 10 years be used for the design of water management and water treatment facilities. Where settlement ponds or other water-retaining facilities are installed, contingency measures should be in place to release water via a spillway or similar should a more serious rainfall event occur.

Key guidance

Hall et al (1993). B14 *Design of flood storage reservoirs, simplified flood estimation method*

HR Wallingford and IH (1981). *Wallingford Procedure for design and analysis of urban storm drainage*

IH (1983). *Flood Studies Supplementary Report 14* *Review of regional growth curves*

IH (1999). *Flood estimation handbook*

18.3.1 Estimating flood flow from small catchments (< 0.5 km²)

Flood flows should be estimated from Table 18.2 below, which provides estimates of mean annual flood in litres/second/hectare based on annual rainfall and soil class. Annual rainfall can be estimated from Figure 18.1, or known site-specific values can be used. Soils are divided into the five classes shown in Table 18.3. Maps showing the distribution of these classes are also available.

Table 18.2 *Mean annual flood peak flow for catchments < 50 ha (litres/second/hectare)*

Soil type	Annual rainfall (mm)					
	< 600	600–800	800–1200	1200–1600	1600–3200	> 3200
1	0.3	0.4	0.6	0.9	1.7	2.4
2	1.4	1.8	2.8	4.1	7.7	10.8
3	2.6	3.4	5.2	7.7	14.4	20.1
4	3.3	4.4	6.7	9.9	18.6	26.0
5	4.2	5.5	8.4	12.4	23.3	32.7

Table 18.3 *Soil classes*

General soil description	Runoff potential	Soil class
Well-drained, sandy, loamy or earthy peat soils	Very low	1
Very permeable soils (eg gravel, sand with shallow groundwater or rock	Low	2
Very fine sands, silts and clays. Permeable soils with shallow groundwater in low-lying areas	Moderate	3
Clayey or loamy soils	High	4
Wet uplands, shallow, rocky soils on steep slopes, peats with impermeable layers at shallow depth	Very high	5

By multiplying flood flow in litres/second/hectare by the catchment area (in hectares), an estimated runoff volume in litres/second can be determined. To calculate the estimated flood for a specific return period, the mean annual flood can be multiplied by a factor for range of return periods. These factors depend on geographical region, but a reasonable estimate for small catchments is provided by the national averages shown in Table 18.4.

Table 18.4 *Factors for different return periods*

Return period (years)	5	10	25	50
Multiplier	1.22	1.48	1.88	2.22

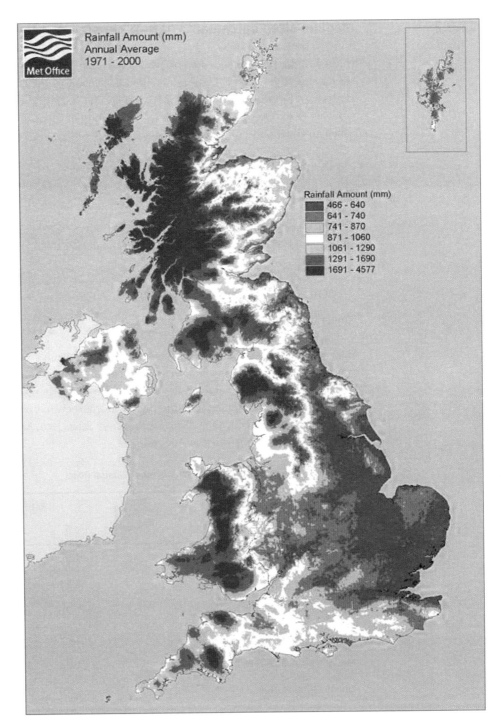

Figure 18.1 *Rainfall amount annual average (mm) 1971–2000 (courtesy Met Office)*

Estimating peak runoff flow on a small catchment (worked example in Appendix 2)

1. Define annual rainfall from Figure 18.1.
2. Define soil class from Table 18.3.
3. Use Table 18.2 to calculate peak flow per hectare.
4. Multiply flow by catchment area in hectares.
5. Multiply by relevant factor for given return period in Table 18.4.

18.3.2 Medium-sized catchments (0.5–25 km²)

In the UK the recommended approach for the calculation of runoff from rural catchments larger than 0.5 km² is a method developed by the Centre for Ecology and Hydrology, formerly the Institute of Hydrology (IH, 1994). The report, which is available free from the Centre, was based on a study of 71 small rural catchments from which an equation to calculate the mean annual flood (Qa) was developed:

$$Qa = 0.00108 \times AREA^{0.89} \times SAAR^{1.17} \times SOIL^{2.17}$$

where:

AREA (km²) is the catchment plan area

SAAR (mm) is the standard average annual rainfall (1961–90)

SOIL is the soil index, defined as:

$$SOIL = (0.15 S_1 + 0.3 S_2 + 0.4 S_3 + 0.45 S_4 + 0.5 S_5) / (S_1 + S_2 + S_3 + S_4 + S_5)$$

Numerical factors are constants.

Maps of SAAR and SOIL are available that enable the mean annual flood to be rapidly assessed. S_1, S_2 etc are soil characteristics derived from mapped data. The mean annual flood is a useful measure of flood flows that have a return period of 2.3 years. It is used as an indicator, which can be scaled to the flow for a required return period, by applying a regionalised scaling factor, as shown in Table 18.5. An example calculation using this method is provided in Appendix 2.

Table 18.5 *Regional factors for scaling mean annual flood*

Region	Return period (years)			
	5	10	25	50
Scottish Highlands	1.20	1.45	1.81	2.12
Lowland Scotland	1.11	1.42	1.81	2.17
North East England	1.25	1.45	1.70	1.90
North West England	1.19	1.38	1.64	1.85
Midlands (Severn-Trent area)	1.23	1.49	1.87	2.20
Lincolnshire, Norfolk	1.29	1.65	2.25	2.83
Southern (inc London and parts of Suffolk, Kent)	1.28	1.62	2.14	2.62
South West England	1.23	1.49	1.84	2.12
Wales	1.21	1.42	1.71	1.94

18.3.3 Large catchments (> 25 km²)

The generally recognised method for the calculation of flood runoff in river catchments is the *Flood estimation handbook* (IH, 1999). This method is suitable for catchments larger than 25 km² and can produce estimates of flood flows for a wide range of probabilities. The handbook is a software tool that requires some expertise to operate. Assistance with producing flood estimates should be sought from suitably qualified consultants or from the Centre for Ecology and Hydrology at Wallingford. This method is therefore most suitable for sites adjacent to or crossing large rivers where the potential risk is large.

18.3.4 Runoff and storage volume

The calculation of runoff and storage volumes for treatment systems is not a simple process and it involves consideration of a number of factors. As discussed above, runoff depends on catchment size, infiltration rates and rainfall. A crude estimation of the volume of runoff from a site to a treatment system, using the sources of rainfall information above, can be given as:

Runoff volume = site area × rainfall total

In calculating this, the site area must include any area, on or off site, that drains to that particular location (rather than the whole site). (However, any upslope runoff from off site should be diverted across or around the site wherever possible; see Section 18.6.8.) This calculation also assumes that all the rainfall becomes runoff and so will result in very large calculated volumes. A factor of permeability (infiltration rates) must therefore be applied to this:

Runoff volume = site area × rainfall total × permeability factor

Infiltration rates vary depending on soil type (Table 18.1) and the site development (vegetation, bare soil, partially constructed roadway etc), so they cannot easily be specified. Typically, an impermeable surface would be 1 and a stripped construction site in the range 0.4–0.75. Applying these would provide volumes within an acceptable range and would indicate an appropriate capacity for a treatment system.

Usually it is necessary to retain the water in a settlement tank or pond for several hours to allow the suspended solids to settle out. Retention time depends on the particle size, disturbance of the water, depth of water, temperature and particle density. The capacity of the treatment system would need to account for this storage time.

One final consideration that presents a challenge to most linear sites is the space available to construct a treatment system (eg settlement pond) of the required size. This issue is discussed further in Chapter 19 and Section 5.3.

18.4 FLOODING

Certain areas may be prone to regular flooding (floodplains, low-lying areas etc), whereas other areas experience unexpected flooding as a result of exceptional levels of rainfall or alteration of the natural or artificial drainage network.

Low-lying areas, dry valleys or channels along the scheme, including access and haul roads, that are liable to flooding should identified, along with rivers, floodplains and wider flood-prone areas (see key guidance; local knowledge is also very useful).

On-site drainage channels should be kept clear and free of blockages to allow the controlled flow of runoff. Where the risk of flooding exists, it is important to monitor weather forecasts and to arrange with the environmental regulator to receive early warning by telephone or fax within designated flood warning areas (see the Floodline websites).

> **Key guidance**
>
> DTLR (2001). Planning Policy Guidance Note 25 *Development and flood risk*, <www.odpm.gov.uk/index.asp?id=1144113>
>
> Floodline, emergency flood services and automatic telephone updates (where available) in the event of severe weather:
>
>   <www.environment-agency.gov.uk/floodline>
>   <www.sepa.org.uk/flooding/index.html>
>   <www.riversagencyni.gov.uk/rivers/floodemergency-whotocontact.htm>
>
> MetBuild Direct service, fax and email forecasts from the Met Office, <www.met-office.gov.uk/construction/mbdirect/index.html>

Where a flood risk exists detailed flood emergency and contingency plans need to be prepared (see Chapter 15), including arrangements to:

- move mobile plant away from flood-affected areas
- make safe any static plant or compounds within flood-prone areas
- evacuate site operatives from flood-affected areas.

> In the event of a flood, taking all reasonable precautions against pollution may provide a defence against prosecution.

18.5 ESTIMATING SEDIMENT GENERATION

Sediment is generated when surface water runoff erodes the soil. Sedimentation occurs when eroded particles transported by water are deposited. Figure 18.2 shows the relationship between velocity and particle size for the three processes performed by water – erosion, transportation and deposition. It can be seen that very high flow velocities are needed to erode both fine clays and coarse particles, and lower velocities for fine sands. The deposition area of the graph does not extend to very fine particles, which, once eroded, are easily transported, even at low runoff velocities, and will not settle out. For coarser particles, the boundary between transportation and deposition is steep, so a slight drop in runoff velocity causes coarser particles to settle out.

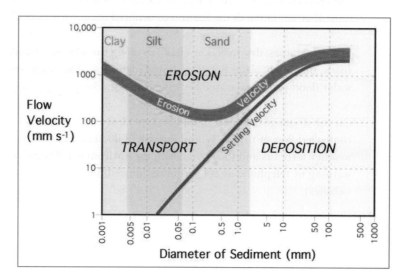

Figure 18.2 *Relationship between stream flow velocity (1 m above bed level) and particle erosion, transport and deposition (courtesy <PhysicalGeography.net>)*

18.5.1 Universal soil loss equation

Numerical models have been developed to estimate sediment generation. The principle factors influencing erosion are the rainfall intensity, soil erodibility, slope gradient, slope length and surface cover (if any). These have been combined within the Revised Universal Soil Loss Equation (RUSLE) to predict soil loss by sheet erosion. This is widely used in the USA, but very little of the required mapped data exists for UK soils. Nevertheless the north-eastern US seaboard resembles many parts of the UK in terms of soil type and climate.

> **Key guidance**
>
> Typical values and worked examples can be obtained from Section 3 (Gaffney and Lake, 2004) of *Draft New York standards and specifications for erosion and sediment control*, <www.dec.state.ny.us/website/dow/toolbox/escstandards/3rusle.pdf>

The equation is as follows:

A = RK(LS)

A is the calculated soil loss per acre per year. It can be scaled to a percentage in any month in the year, with around 50 per cent of the annual soil erosion occurring in the three months between July and September in the north-eastern seaboard of the USA (Gaffney and Lake, 2003).

R is the rainfall value reflecting rainfall energy multiplied by the intensity factor. On the NE seaboard, the R value varies from 75 to 200 (Gaffney and Lake, 2003), so, applying an R value of 100 to the UK would give an estimation within the range of -30 to +200 per cent.

K is the soil erodibility factor base on soil type. K values vary from around 0.2 for gravely sand, to 0.6 for silty clay, but are currently determined by geography rather than geotechnically.

LS is the horizontal slope length (**L**), in feet, multiplied by the slope gradient (**S**).

The equation can also assess the effectiveness of sediment control measures:

A = RK(LS)CP

C is a factor reflecting the cover over the soil surface, such as mulch, geotextile etc (see Sections 18.6.1–18.6.5) and is an excellent indicator of the benefit of covering soil surfaces. C values for various stages of vegetation establishment can also be used. The C value for exposed soil on site would be 1. The C value for an application of woodchips on a 20 per cent slope is 0.08, thus reducing the calculated eroded soil by 92 per cent!

P is a factor representing management options on construction sites, also clearly indicating which soil surfaces erode the most (see Section 18.6.3). Bare, loose soil has a P factor of 1. A roughened surface is 0.9 (*decreasing* eroded soil by 10 per cent), a compact, smoothed surface is 1.2 (*increasing* eroded soil by 20 per cent) and furrowed surfaces from 0.5 to 0.8 depending on slope gradient.

18.6 EROSION AND SEDIMENT CONTROL MEASURES

The effectiveness of any sediment control measure depends upon weather conditions, site characteristics and construction activities.

1. Rainfall event – the amount of sediment generated depends on the conditions prior to rainfall and the intensity and duration of the event itself.

2. Site characteristics – as the site changes during construction, the area contributing to sediment generation varies. For example on a road scheme, cleared areas increase (more exposed soil), followed by an increase in impermeable areas (less exposed soil, but increased runoff, probably into new drainage).

3. Soil type – different sediment control measures are more capable of trapping different soil particle sizes.

4. Method of sediment control – different methods perform at different efficiencies.

5. Adequacy of sediment control measures – for any one or combination of sediment control measures, a continual review of effectiveness, maintenance and modification of control measures is required throughout the construction to address variable and changing circumstances.

18.6.1 Existing vegetation

Mature vegetation has extensive root systems that help to hold soil in place and prevent soil from drying rapidly, thus reducing erosion. Existing vegetation is far more effective at preventing erosion than newly planted vegetation.

The first priority is to minimise the area of exposed ground. The first construction activity on most sites is the clearance of vegetation and topsoil strip. Polluted runoff can be greatly reduced by minimising the time between removal of the vegetation cover and establishment of the post-construction cover. Due to the inherent large scale of linear projects, when programming for works, the site should be divided into phases.

Where no construction activity occurs, or where natural vegetation should be preserved, such as near watercourses, in wooded areas and on protected sites, preservation of existing vegetation should be carefully planned. For further information see BS 5837:1991 *Trees in relation to construction. Recommendations* and also CIRIA B10 *Use of vegetation in civil engineering* (Coppin and Richards, 1990).

> **Vegetation and topsoil strip**
> 1. Leave as much existing vegetation as possible and specify proper care of this vegetation during construction, clearly protecting it with fencing, signs etc.
> 2. Leave existing (or plant new) vegetation along the perimeter of the site, haul roads or stockpiles to provide an effective buffer against sediment leaving the area.
> 3. Leave a 5 m grassed strip next to river banks when stripping topsoil or place grassed soil bunds along river banks etc to prevent site runoff gaining direct access to watercourses.
> 4. Delay clearing and topsoil strip of each phase of works until shortly before construction begins rather than stripping the whole site many months before construction.
> 5. Seed cut and fill slopes as the work progresses.
> 6. Close and stabilise (eg revegetate) open trenches as soon as possible. Sequence projects so that most open portions of the trench are closed before new trenching is begun.

18.6.2 Seeding and planting

Seeding with grasses and ground covers contributes to long-term stabilisation of soil (Figure 18.3). For temporary stabilisation, grasses can be planted, provided sufficient growing season is available before the onset of winter. Temporary stockpiles should be seeded, along with any cleared areas where construction activities have ceased, especially if they have steep sides that might undergo gullying. Seed species should be of local provenance, and ideally the same as those specified for the final landscape scheme. Additionally, diversion drains (see Section 18.6.8) and bare stream banks may require seeding and planting.

> Apply seeding and planting to all graded and cleared areas as soon as construction activities in that area cease. If there is no time to establish grass cover on a slope, covering the slope with woodchips or similar mulch (Section 18.6.4), or even roughening the surface (Section 18.6.3), will provide better erosion control than leaving the slope smooth and uncovered. Other alternatives include the use of geotextiles (Section 18.6.5).

The type of vegetation, preparation, planting time, fertilisation and water requirements should be considered for each application. A maintenance plan, including irrigation and mowing, should be developed to ensure rapid colonisation and growth. The use of fertiliser should be avoided; if used, it should be applied with care, as runoff may carry it into watercourses.

Figure 18.3 *Before and after seeding: erosion gulleys on non-vegetated slope (courtesy IECA)*

Seeding with grasses may not be effective on steep (greater than 2:1) slopes, waterlogged areas or on compacted or poor soils. Geotextile fabrics (see Section 18.6.5), or hydraulic seeding, in which an appropriate binder is incorporated in the application, may be required on steeper slopes. Seeding may have to be combined with mulching (see Section 18.6.4) to ensure effective sediment control during the initial growth phase.

18.6.3 Slope roughening and benching

Graded areas with smooth, hard surfaces give a false impression of "finished grading" and a job well done. Rough slope surfaces with uneven soil and rocks left in place may appear unattractive or unfinished at first, but they encourage water infiltration, speed the establishment of vegetation and decrease runoff velocity. If there is no time to establish grass cover on a slope, roughening the slope surface will provide better erosion control than if the slope is left smooth. The effectiveness of these measures is demonstrated in Section 18.5.1.

There are different methods for achieving this on a slope. Factors to be considered in choosing a method are slope steepness, geotechnical ground conditions, mowing requirements, and whether the slope is formed by cutting or filling. Geotechnical advice should be sought to determine whether the increased infiltration encouraged by a rough surface may induce instability in high or steep slopes.

Benching is particularly appropriate for major earthworks in soils containing large amounts of soft rock such as shale. Graded areas steeper than 3:1 should be benched to help vegetation become attached and also trap soil eroded from the slopes above.

Roughening can be achieved by using a tracked vehicle to leave track imprints parallel to the slope or, with an attachment, to create grooves across the slope (Figure 18.4). Slopes that will be mowed (which should have slopes less than 3:1) may have small furrows left by disking, harrowing, raking or seed-planting machinery operated on the contour.

 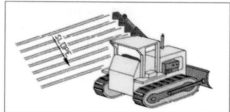

Figure 18.4 *Tracking (left) and furrowing (right) (Oregon DEQ, 2004)*

18.6.4 Mulching and binders

Mulches include organic materials, straw, wood chips, bark or other wood fibres and gravel and are used to stabilise or protect cleared or seeded areas, either temporarily or permanently. Mulching can reduce erosion by between 20 and 95 per cent, depending on the slope gradient, soil type and mulch material (Oregon DEQ, 2004), as demonstrated in Section 18.5.1.

Mulch prevents erosion by protecting the soil surface and holding seeds, fertilisers and topsoil in place until growth occurs. It may be used with netting to supplement soil stabilisation. The supplier or contractor should recommend the type of mulch and binders and their application rates.

Binders are biodegradable adhesives that can be applied directly to the soil or over a layer of mulch. They are usually mixed with water and sprayed on. They are also used to reduce dust generation from stockpiles, haul roads and compounds. Seek advice from a supplier to select the appropriate binder and application method.

The maintenance regime should include inspections weekly and after rain for areas where mulch has blown away or been pushed together. Organic mulches are not permanent erosion control measures.

18.6.5 Geotextiles and mats

Meshes, netting, mats and sheeting made of natural or synthetic material can be used to stabilise soil temporarily or permanently. Typically they are suited to post-construction site stabilisation, but they may be used for temporary stabilisation of easily eroded soils in sensitive areas, including channels and streams where flow velocity may cause erosion.

Matting may be applied to disturbed soils and places where existing vegetation has been removed. Organic matting materials, such as jute or straw, provide temporary protection until permanent vegetation is established and will not need to be removed, as they will rot down. They may also be appropriate when seasonal circumstances dictate the need for temporary stabilisation until weather or construction delays are resolved.

Synthetic matting may be used for either temporary or post-construction stabilisation, both with and without vegetation and will be much longer-lasting. Proper installation of matting is critical in order to obtain firm continuous contact with the soil where revegetation is required (Figure 18.5).

Design – geotextile and matting
- apply by rolling down slope of in the direction of water flow
- overlap both edges
- make sure products are securely staked down
- where impermeable mats are used (eg plastic sheet), provide a straw or rock barrier or silt fence at the toe of the slope.

Figure 18.5 *Coconut matting and silt fence (Oregon DEQ, 2004)*

18.6.6 Haul roads and site access

Access roads, parking areas and haul roads need to be stabilised immediately after grading and maintained thereafter. During wet weather they often generate significant quantities of sediment that may pollute nearby streams or be transported off site on the wheels of construction vehicles.

Efficient haul road stabilisation not only reduces on-site erosion but can significantly speed on-site work, avoid instances of immobilised machinery and delivery vehicles, and generally improve site efficiency and working conditions during adverse weather.

Figure 18.6 (left)
Haul road (courtesy Transco)

Figure 18.7 (above)
Haul route runoff (courtesy Transco)

Ditches should be constructed on either side, or on the downslope side, of haul roads to channel water to a treatment area (settlement etc); see Section 18.6.8. On sloping ground, runoff will collect at the lowest point – earth bunds along the length, or a cut-off trench at the lowest point, can be used to direct runoff away to a suitable area for treatment (see case study in Section 20.8).

If plant or vehicles have to make repeated crossings of a watercourse, a temporary haul road bridge or flumed/culverted crossing should be erected. Straw bales or sand bags can be placed along the sides of temporary or existing bridges to prevent runoff into the water below (see Section 20.3).

> **Design – haul roads**
>
> - Follow the contour of the natural terrain to the maximum extent possible.
> - Slope should not exceed 15 per cent.
> - Grade haul roads to drain across rather than along (ie the shortest distance) or construct low earth bunds to divert runoff off the road (see case study in Section 20.9).
> - Provide drainage ditches (see Section 18.6.8) on each side of the haul road (or on down-slope side). Simple gravel bunds without a trench can also be used to filter road runoff.

Site entrances need particular attention, as sediment tracked off site on to public roads can pollute surface water drains. A stabilised construction entrance (see Figure 18.8), consisting of a pad of aggregate underlain with geotextile fabric, should be located at any point where traffic will enter or leave a construction site to or from a public right of way. The stabilised entrance should extend for at least 15 m from the paved road and consists of a minimum of 150 mm of gravel, laid on a geotextile fabric where soft ground conditions dictate.

Stabilised construction entrances are moderately effective in removing sediment from equipment leaving a construction site. More formal wheel-washing facilities may be required where particularly dirty vehicles, such as haulage trucks, are regularly leaving the site.

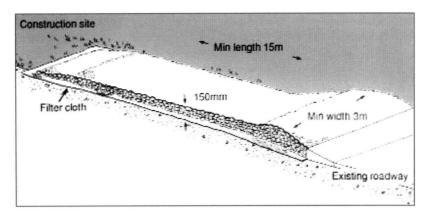

Figure 18.8 *Stabilised construction entrance; geotextile would be used as "filter cloth" in the UK (ACT, 1998)*

18.6.7 Protection of stockpiles

Stockpiles can be a significant source of erosion and sediment. To minimise the loss of sediment from stockpiles they should be:

- located away from drains and watercourses
- seeded or provided with other stabilisation measures appropriate to the length of time stored
- provided with earth bunds or another form of diversion to keep runoff away from stockpiles
- provided with silt fences or straw/rock barriers at the toe of the stockpile to mitigate runoff during rain events.

18.6.8 Diversion drains

Diversion drains are simple linear ditches, often with an earth bund, for channelling water to a desired location (Figures 18.9 and 18.10). Diversion drains should be used:

- for diverting upslope runoff, particularly off-site runoff, along, across or around the site
- for collecting and channelling silty runoff downslope of the site to prevent it leaving the site
- around the toe of stockpiles or cut/fill embankments
- at the top of slopes, channelling runoff to a slope drain (see Section 18.6.10)
- around any other disturbed area
- as a containment measure around contaminated soil or other materials.

Where there is clean water running from above or across the site that picks up sediment from the site, consideration should be given to piping the water across the site or using diversion ditches. This will minimise the runoff that is of concern on the construction site. Where runoff volumes are small, an earth perimeter bund may be sufficient to channel water.

If the drains themselves are being eroded, they can be lined with geotextile fabric (Figure 18.10), or large stones/boulders or check dams (Section 18.6.11) can be emplaced at regular intervals to reduce velocity and increase infiltration. Where long-term drains are vegetated, they are often referred to as "swales", although these tend to be used in SUDS for permanent works.

The outflow of the drain should be directed to a stabilised area known as a "level spreader" or to a settlement/balancing pond.

Figure 18.9 Left: *diversion drain alongside sloping pipeline easement (courtesy Alfred McAlpine)*

Figure 18.10 Above: *diversion drain (lined with geotextile) (ACT, 1998)*

Design – diversion drains and bunds

- Temporary drains or any other diversion of runoff should not adversely impact upstream or downstream properties.
- No more than 2 ha may drain to a temporary drain.
- Place the drain above, not on, a cut and fill slope.
- Drain bottom width should be at least 0.6 m.
- Depth of the drain should be at least 0.5 m.
- Side slopes should be 2:1 or flatter.
- Drain should be laid at a grade of at least 1 per cent, but not more than 15 per cent.
- The drain must not be overtopped by a 10-year, 24-hour storm.
- Do not construct bunds from soils that may be easily eroded.
- Compact bunds by earth-moving equipment to prevent failure in a storm.
- Do not operate construction vehicles across a drain or bund unless a stabilised crossing is provided; the bund top width may be wider and side slopes may be flatter at crossings.
- Stabilise all ditches where erosion is likely.
- Any drain that conveys sediment-laden runoff must be diverted into a settlement pond or provided with some other treatment before the water is discharged from the site.

18.6.9 Level spreader

The purpose of a level spreader (Figure 18.11) is to convert a concentrated flow of sediment-free runoff (eg from a drainage ditch) into sheet flow and to discharge it on to flat-lying, vegetated ground without causing erosion. It is used to terminate a diversion drain or perimeter earth bund (Section 18.6.8) or to distribute spill flow from a retention pond.

The level spreader is only suitable on undisturbed vegetated areas and where the water will not be reconcentrated into a flow channel immediately below the point of discharge (ie on level ground rather than a slope).

The design criteria for level spreader require that it has a maximum of 1 m of length per 0.1 m³ peak rate of flow from a 10-year frequency rainfall event. Sill length depends on contributing catchment, slope and ground conditions. In any case, a minimum length of 4 m and a maximum length of 25 m are recommended (Figure 18.11).

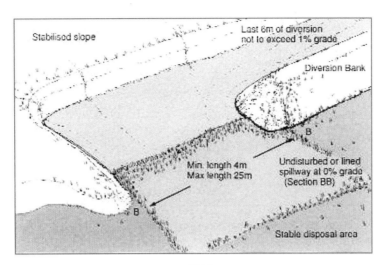

Figure 18.11 *Level spreader (ACT, 1998)*

18.6.10 Slope drains

A slope drain is a temporary pipe or lined channel to drain the top of a slope to a stable discharge point at the bottom of a slope without causing erosion (Figures 18.12 and 18.13). It prevents runoff from flowing directly down the slope by confining all the runoff into an enclosed pipe or channel. Temporary slope drains are used in conjunction with bunds along the top edges of slopes that direct water to an inlet at the top of the drain.

Figure 18.12 *Slope drain (Oregon DEQ, 2004)*

Design – slope drains

- Provide a secure inlet surrounded with bunds to prevent gully erosion.
- Use rigid or flexible pipe, or line a drainage way with plastic sheet or geotextile matting.
- Anchor the pipe or sheeting to the slope.
- Design to convey at least the peak of a 10-year, 24-hour storm.
- Each slope drain should service no more than 2 ha.
- Design the outlet to prevent erosion from the high exit velocity, using energy dissipaters (straw bale, pile of rock etc).
- Direct drainage into a stable sediment trap, diversion drain or settlement pond.

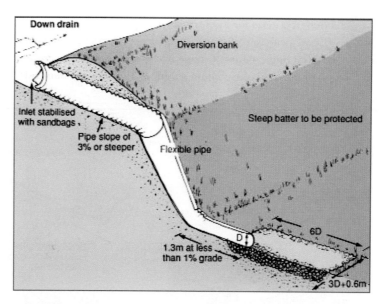

Figure 18.13 *Slope drain detail (ACT, 1998)*

18.6.11 Check dams and sediment traps

These are small temporary dams constructed across a swale or drainage ditch to reduce the velocity of concentrated runoff, thereby reducing erosion of the swale or ditch, and promoting sedimentation behind the dam. If properly anchored, wood, straw bales, hay bales or rock filter bunds may be used for check dams (Figure 18.14).

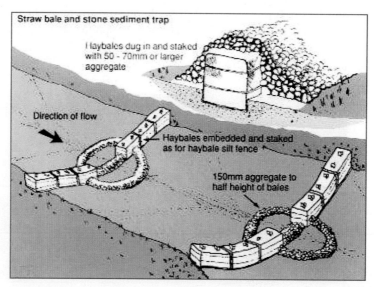

Figure 18.14 *Check dams; straw bales can be used in place of hay bales (ACT, 1998)*

> **Design – check dams**
>
> - Do not install in natural watercourses or channels.
> - Place at a distance and height to allow small pools to form between each one.
> - The bottom elevation of the upper dam should ideally be at the same as the top elevation of the next dam downstream.
> - The centre should be lower than the edges to allow flood runoff to overtop the dams rather than cause upstream flooding or scouring around the dams.
> - Embed straw bales and logs, where used, to prevent scour.
> - Inspect regularly and remove sediment accumulation when it reaches half the dam's height.

18.6.12 Silt fence

A silt fence comprises a geotextile filter fabric, straw bales or a combination of the two installed in the path of sheet flow runoff to filter out heavy sediments (Figure 18.15). At best, a silt fence will remove heavy settleable solids, but it is not effective at reducing turbidity (fine solids in suspension) (see Table 18.6).

Table 18.6 *Range of control for silt fences*

Sand	80–90 per cent
Silt-loam	50–80 per cent
Silt-clay-loam	0–20 per cent

Silt fences are suitable for:

- use along the perimeter of the site
- below the toe of a cleared slope
- around temporary stockpile areas
- across swales with catchments less than 0.5 ha.

The silt fence detains sediment-laden water, promoting sedimentation behind the fence. Posts support the filter fabric, the base of which should be well buried in the ground. Commercially available products permit easy installation and use; the correct method of installation should be ascertained from the supplier. Selection of a filter fabric is dependent on soil conditions at the construction site – the filter fabric should retain the soil yet have openings large enough to permit drainage and avoid clogging. Accumulated sediments need to be cleared away regularly (commercially available fences have a printed indicator line for maximum sediment level), and the fence should be replaced if it rips, sags or is pulled out of the ground.

Figure 18.15 *Fabric silt fence at toe of stockpile (courtesy IECA)*

> **Design – silt fence** (follow supplier guidelines when installing commercial products)
>
> - Use in areas where sheet flow occurs.
> - Install along a level contour so water does not pond more than 400 mm at any point.
> - Leave an area behind the fence for runoff to pond and sediment to settle.
> - Make sure no more than 0.5 ha of concentrated flow drains to any point along the silt fence.
> - Fix the fabric to strong supporting posts at regular intervals.
> - Bury the filter fabric at least 100 mm into the ground (commercially produced fences incorporate a printed line indicating the correct installation depth).
> - Turn the ends of fence uphill to prevent runoff flowing around the end of the fence.
> - Support the fence by a wire mesh if the fabric selected does not have sufficient strength.
> - Clear away accumulated silt regularly; commercially produced silt fences have a printed indicator line over which silt should not accumulate.

Straw bales can also be used to filter out heavy sediments. During wet weather bales deteriorate rapidly and require frequent replacement, but they are a cost-effective temporary measure. If straw bales are merely placed on the ground without proper anchoring and trenching (Figure 18.16), they will provide only minimal erosion control.

Greater efficiency can be achieved by wrapping straw bales in geotextile (see Figure 18.17) or combining a silt fence with straw bales (Figure 18.18).

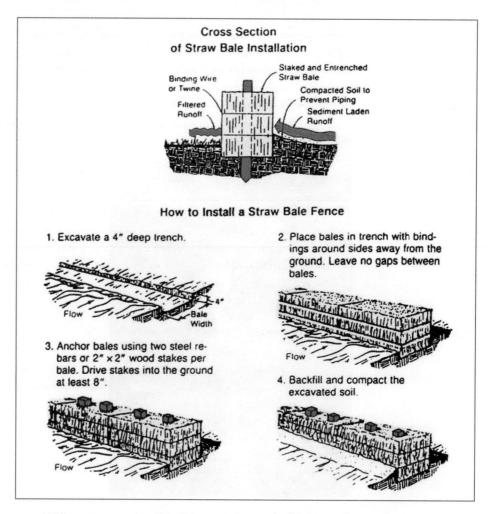

Figure 18.16 *Straw bale installation (courtesy Oregon DEQ, 2004)*

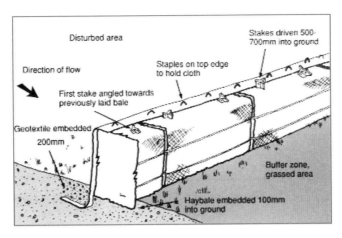

Figure 18.17 *Straw bale (or hay bale) and geotextile fence installation (ACT, 1998)*

Figure 18.18 *Silt fence and straw bale arrangement to control silty runoff; note buffer zone between fence and earthworks (courtesy Mowlem)*

18.7 PROTECTING EXISTING AND PRE-CONSTRUCTION DRAINAGE

A drainage plan for the site (see Chapter 12) should be prepared that identifies the location of all existing surface water drains, sewers, manholes, field drains and pre-construction drainage. A linear route may intersect a drainage pipe a number of times, so the activities at one location may affect the drainage at other locations (perhaps in the hands of other contractors) along the route. Greater, cumulative effects may in turn be experienced farther downstream.

18.7.1 Storm drains

Stormwater runoff is of particular concern along routes passing through urban or built-up areas, or following existing infrastructure (highway widening, for example). When working on existing drainage systems, runoff needs to be prevented from entering open drains or gulleys. Temporary measures can be put in place at the outfall (or intersection with other drainage) to remove sediments and oil, such as a catch pit, sump or a geotextile screen, or the pipe/culvert can be temporarily blocked or diverted.

Where an existing surface water sewer system is present on or adjacent to the site (eg highway or railway drainage), inlets to the system should be protected by a similar method to those illustrated in Figures 18.19 and 18.20.

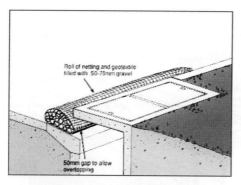

Figure 18.19 *Kerb inlet control (ACT, 1998)*   **Figure 18.20** *Storm drain control (ACT, 1998)*

> **Construction company fined £2000 for polluting brook from site drainage** (source: EA press release, Dec 2004)
>
> A construction company has been fined £2000 and ordered to pay £795 court costs after polluting a tributary of the River Colne. The company admitted causing construction runoff containing high concentrations of suspended solids to enter Wessenden Brook. The EA found muddy brown water in Wessenden Brook, which was coming from a pipe below the road bridge.
>
> The pollution stretched for an estimated 120 m along Wessenden Brook until it joined up with the River Colne. Officers traced the muddy water to a large construction site where they found the water seeping into drains. The court heard how the Environment Agency had, on several occasions, reason to speak to the company about contamination flowing from its site into the watercourses prior to the incident. In its defence, the company maintained that it took these matters seriously and had placed bunding along the road to prevent contaminated water running off.
>
> The Agency's environment officer said: "The construction industry is responsible for a significant number of pollution incidents every year". In this case the construction company had a difficult site to work on because the land was on a slope, but this demonstrates how companies should have good working practices in place early on to prevent problems from occurring.

18.7.2 Land drains

In rural areas, land (field) drains are likely to be encountered. Existing land drains can be difficult to detect since they are often constructed of local stone and blend into the environment (Figures 18.21 and 18.22). During heavy rain land drains can transport silt pollution from the site into watercourses (Figure 18.23). Existing land drains should be identified and the outfalls located and monitored. It is important to be able to isolate land drainage if necessary. During construction newly installed drains should not drain working areas stripped of topsoil. Where land drainage may present a pollution risk, solid (not perforated) pipe should be used and in-line filters and sumps installed. The filter medium should not be brought up to ground level (see also "Pre-earthworks drainage" in Section 18.7.3).

Where land drains are truncated by the scheme they can discharge large volumes of water on to the site. If they were simply sealed, land upslope would become waterlogged and flooded, so it is important to intercept and divert flows (with permission) across the site to a suitable outfall downslope in a controlled manner.

The main contractor can also be held responsible for the quality of water diverted through the works and discharged from an outfall used during construction. It is important, therefore, to be aware of any activities upstream such as muck-spreading or ploughing that may cause polluted water to enter the diverted land drains.

Figure 18.21
Above: *land drain uncovered during construction work (courtesy Transco)*

Figure 18.22
Above right: *land drains often blend in to the local environment so can be hard to identify (courtesy Transco)*

Figure 18.23
Right: *silty water from a land drain entering a watercourse (courtesy Transco)*

> If you observe a polluting activity upstream that may affect the water quality of site drainage or outfall discharge, report it immediately to the environmental regulator.
>
> Reporting pollution reduces the risk of prosecution and can be used in mitigation should there be a prosecution.

18.7.3 Pre-earthworks drainage

It is common practice to install pre-earthworks drains on linear schemes. These permanently intercept field drains and groundwater, and drain water away from the working area, usually to a watercourse. Typically, pre-earthworks drains consist of drainage coil installed in the base of a trench backfilled with drainage stone up to the surface.

In some circumstances, silt-contaminated runoff enters pre-earthworks drains via the stone backfill. The drains carry the silt-laden water to the watercourse without any treatment and can cause pollution. The surface of the pre-earthworks drains should be sealed with clay to prevent ingress from silty runoff. The runoff can be controlled by one of the methods suggested in this chapter.

Other temporary "pre-construction" drainage may take the form of diversion ditches (see Section 18.6.8) or slope drains (see Section 18.6.10).

18.8 SUSTAINABLE DRAINAGE SYSTEMS (SUDS)

Many sites are being designed to incorporate the SUDS approach to drainage to attenuate and treat runoff during their operational phase, for example using French drains, swales, ponds and wetlands. SUDS may also be part of the temporary drainage system planned to treat runoff during construction. The interaction between these two systems needs careful consideration. It may be possible to combine some parts of the two systems, for instance by using final retention ponds during the construction phase, but in other cases the sediment generated during the construction phase can destroy the effectiveness of the planned SUDS system.

> Sediment volumes are much greater during construction than permanent SUDS are designed for.

If it is possible to use part or all of the final SUDS system during the construction phase then sediment needs to be cleaned out of ponds etc after the construction phase is completed. Sediment volumes are much greater during construction than the permanent systems are designed for, which could reduce the life of the SUDS. There is also a greater risk of contaminated sediment (including cementitious material) being in the sediment, which could harm aquatic life in the SUDS.

If the final SUDS system will not be part of the sediment control system during construction then great care needs to be taken to ensure sediment does not enter any part of the SUDS. The main sources to consider are site runoff and the tracking of site vehicles over completed surfaces. Infiltration systems, particularly permeable pavement, are especially prone to clogging by excessive sediment and these may be become contaminated by site runoff or by mud deposited during site works. For this reason, installation of such systems is best done at the very end of the construction phase.

19 Water treatment methods and disposal

19.1 INTRODUCTION

Linear sites are not well suited to some of the traditional methods of water treatment because they generate large volumes of runoff, it is difficult to divert surface water around the site and space for treatment facilities is limited.

In all cases, the first priority should be to control the source of pollution by:

- controlling **runoff** from the site, which will otherwise erode exposed soil, haul roads and stockpiles (follow the guidance in Chapter 18)
- **dewatering** excavations in a controlled manner (follow the guidance in Chapter 21)
- using best practice methodologies when **working in or near water** (follow the guidance in Chapter 20)
- storing and using **fuels and oils** according to best practice to avoid leaks and spills (follow the guidance in Chapter 16)
- controlling **concrete and grouting** operations (follow the guidance in Chapter 22).

> Sediment volumes are much greater during construction than permanent SUDS are designed for.

This chapter provides guidance on the options available for treating water to remove common pollutants, depending on the volume of water and the space available as well as guidance on how and where water can be disposed of without causing pollution.

The most common pollutants present in water on site are:

- sediment (suspended solids such as soil, mud, silt and clay)
- cement products
- hydrocarbons such as fuel and oil
- heavy metals
- organic wastes (sewage and effluent from welfare facilities).

19.2 SEDIMENT

Sediment, including all suspended solids, silt clay, mud, sand etc, is the single main pollutant generated at construction sites. Once all source control measures have been put in place (see Chapter 18), several options are available for removing sediment from collected water. More than one treatment method may have to be used, depending on site characteristics – eg volume of water, location of discharge. To establish the best approach for a site, the following options need to be assessed.

1. Pump to grassland, soakaway or infiltration basin.
2. Pump to an adequately sized settlement facility (tank, lagoon etc).
3. Pass through a silt trap or filtration system.
4. Install specialist treatment equipment.
5. Pump into a tanker and dispose off site.

> Oil, cement, concrete and other chemicals will stick to fine sediments, causing further pollution.

19.2.1 Pumping to grassed land – infiltration

This is effectively a treatment-and-disposal solution, but it is not appropriate for long-term use. This option is suitable for water that is unpolluted aside from its silt content. The performance of such methods depends on the infiltration and permeability of the underlying ground (see Table 19.1). Saturated ground (following heavy rainfall), high water table, shallow impermeable layers (clay, rockhead), protected groundwater, contaminated ground or large inflow of water will prevent this being a suitable option. The silty water should contain no chemical or biological pollutants such as oil, concrete or sewage. Before pumping, permission should be sought from the landowner. Approval may also be required from the environmental regulator if the area has sensitive groundwater.

Table 19.1 *Typical infiltration rates for various soils*

Soil texture		Sand	Sandy loam	Loam	Clay loam	Silty clay	Clay
Typical	mm/h (range)	50 (25–250)	25 (13–75)	13 (8–20)	8 (3–15)	3 (0.3–5)	5 (1–10)
Typical	l/min/ha*	8000	4000	2000	1300	500	800

* Assuming the water to be spread evenly over the surface (Chant-Hall *et al*, 2005a).

Infiltration areas should be well away from main site excavations to avoid recirculation through the ground. If space is a limiting factor, it may be possible to channel water to part of the site that has not yet been stripped of topsoil and vegetation, or to gain permission to use adjacent land in rural areas.

> Always check for the presence of land drains – any water infiltrating the ground will flow into land drains and discharge directly to a watercourse (see Figure 19.1 and the following case study).

Figure 19.1
Silty water discharging through land drain into ditch (courtesy Transco)

> **Silt pollution via land drains costs contractors £30 000** (source: *The ENDS Report*, Mar 2003)
>
> A construction company and an engineering company have each been fined £15 000 after rivers in north-east Scotland were repeatedly polluted with silt from a construction project. More than 50 000 fish were killed and one incident turned 20 km of a burn brown with silt.
>
> The two companies had formed a joint venture to construct a new gas pipeline. Before the project began in early 2001, SEPA advised the company and its contractors that they must take adequate steps to prevent water pollution. The contractor's own pre-construction environmental statement recognised the River Ythan as "highly sensitive". The court heard how SEPA began to receive complaints from the local fisheries board in the summer of 2001. High levels of silt affected several rivers and burns in the Ythan catchment. The Ythan is a site of special scientific interest and a national nature reserve.
>
> SEPA traced the pollution to the pipeline construction work. In sinking the pipeline, silt-laden rainwater had to be pumped out of the workings. Instead of being disposed of safely, the contractors were pumping it on to surrounding fields. The muddy water entered watercourses through field drains at several sites. In one incident, a tanker brought in by the contractors to remove the runoff was allowed to overflow, polluting the Ebrie burn.
>
> SEPA said that the incidents continued despite numerous meetings. The Agency repeatedly warned the companies that they were breaking the law and needed to put in place pollution prevention measures. A series of events led to the deaths of more than 50 000 fish at a fish farm downstream of the operations. Samples taken by SEPA from some of the polluted waters, showed silt levels 450 times above normal background levels in the Ythan.

Pumping water directly on to a grassy area is most suitable for small volumes pumped at intervals, for example dewatering an excavation at the start of each shift or dewatering a trench with slow water ingress. It should not be done over extended periods otherwise the soil will become waterlogged and the grass clogged with sediment.

Waterlogging can be avoided by moving the pipe outlet regularly or by using an irrigation spray or "rainmaker" (see following case study) to spread water. Pumping needs to be at a rate that allows water to infiltrate (see Table 19.1 and key guidance). A typical agricultural sprayer discharges at about 1000 l/min.

An infiltration basin or soakaway can be established. These work in a similar way to settlement ponds, but have no outfall to a watercourse and rely on infiltration of water into the underlying ground.

The pond should be sized according to the inflow (see Chapter 18 for guidance on runoff calculations) and the ground permeability (see Table 19.1 and key guidance). Infiltration basins are sometimes specified for the permanent works. If this is the case, provided they are of the correct capacity and can be cleaned out if necessary on project completion, they can be built at an early stage for use during construction.

> **Infiltration**
>
> 1. Check treatment option – infiltration is not suitable for oil, concrete and chemical contamination.
> 2. Not suitable where there is wet ground, land drains (unless they can be sealed or diverted), shallow water table or protected groundwater areas.
> 3. Before pumping, seek permission from the landowner. Approval may also be required from the environmental regulator if the area has sensitive groundwater.
> 4. Pump at a rate that will allow water to infiltrate the ground (see table above).
> 5. Make sure water does not run overland and into nearby watercourses.
> 6. Move the outfall hose from time to time or use an agricultural rainmaker to avoid saturation of one area, generation of a plume of silt and smothering of vegetation.
> 7. For larger volumes over a longer timeframe consider establishing an infiltration basin.

Key guidance

Infiltration calculations and design

Bettess (1996). R156 *Infiltration drainage – manual of good practice*

Highways Agency (2001). HA103/01 "Vegetative treatment systems for highway runoff", *DMRB*, Sec 4.2.1

BRE (1991). Digest 365 *Soakaway design*

Martin et al (2000a). C521 *Sustainable urban drainage systems – design manual for Scotland and Northern Ireland*

Martin et al (2000b). C522 *Sustainable urban drainage systems – design manual for England and Wales*

Masters-Williams et al (2001). C532 *Control of water pollution from construction sites*

Case study environmental briefing – rain guns (source: Mowlem)

Mowlem Civil Engineering — Environmental Briefing – 2004 006

The use of 'Rain Guns' for the Disposal of Silty Water

Agricultural irrigation sprays or 'Rain Guns' have been used during the Otterburn Training Area Development project as an alternative solution for dealing with silty water from excavations and culvert works. This circular outlines the implications of this method.

Reason for use: Settlement and filtration methods could not provide the retention time necessary to remove the fine boulder clay sediment from the discharge. Increased volumes of water infiltration into the excavation compounded the problem. A 'Rain Gun' was successfully used to compliment existing silt measures and provide extra silty water handling capacity.

Advantages: The use of a Rain Gun provides a comparatively high handling capacity for silty water as it doesn't rely on retention time or filtration. It also avoids use of flocculating agents or extra settlement facilities and associated cost implications. Finally it can provide irrigation for dry areas.

Conditions for use: This method needs a large vegetated broadcast area and good quality pumped feed due to the pressures involved. An automatic tracking system is required on the gun to ensure a varied discharge across the broadcast area and prevent local saturation or erosion. The disposal of silty water in this manner requires prior consent from both landowner and Environment Agency.

Potential problems to be aware of:
- The system works under pressure so bagging condition and fitting is important - poor installation can lead to the uncontrolled escape of silty water.
- The broadcast area should be checked for drainage grips or gullies as these can cause direct silty water runoff.
- The broadcast area can become saturated. This could mean only sporadic pumping or regular movement of Rain Gun.
- The jet can block but this risk can be reduced by installing a stone filter to sump of excavation and protecting the pump draw with perforated piping as shown opposite.
- Frost can damage working parts of the jet so the system should be drained in frosty conditions.

More information on the use of Rain Guns can be obtained from
David Mason (01344 742238) or Chris Hillman (01830 520171)

19.2.2 Settlement ponds

A settlement pond, or lagoon, is one of the simplest and most effective treatment methods available and requires less maintenance than other sediment control techniques. Site runoff or water pumped from excavations is channelled into a pond constructed specifically to allow any suspended solids to settle out before discharge. A pond is highly effective in attenuating storm flows and containing water for quality monitoring or any other required treatment.

Settlement ponds may not be practicable for all linear sites, as they can require a relatively large land take. Ideally, suitable areas of land should be identified at an early stage in the project (see Section 5.3). Where such land is not available, and the site is restricted, the use of settlement tanks for smaller volumes (Section 19.2.3) and the addition of flocculants for larger volumes (Section 19.2.6) can help hasten settlement before discharge.

Certain projects, particularly highway schemes, include balancing ponds as part of the permanent works. These should be constructed as early in the scheme as possible so they can be used as settlement ponds throughout construction to control and treat surface water. The pond should be handed over to the client in a suitable condition at the end of the construction phase, and may need to be dredged to remove silts.

> Where a balancing pond forms part of the permanent works, construct it at the outset and use it to control and treat site water.

Design criteria

Ponds or lagoons need to be designed for stormwater/flood events appropriate to the length of the project. A 1-in-10-year storm event is recommended for temporary works. However, contingency measures should be in place to release water via a spillway or similar in the event of a more serious rainfall event. The guidance in Section 18.3.4 can be used for calculating the volume of runoff that can be anticipated and details on sizing settlement facilities are given in Section 19.2.4. The outfall should be placed as far away from the inlet(s) as possible. Illustrated designs are provided in CIRIA C532 *Control of water pollution from construction sites* (Masters-Williams *et al*, 2001). Figures 19.2–19.4 illustrate examples of typical settlement ponds.

> When designing a settlement pond, select a probability of rainfall that is appropriate to the length of project and the risk of failure. A likely minimum design period for temporary works is **once in 10 years**, with an overspill contingency for greater events.

Depending on the local geology, the pond may require lining (with clay or a proprietary liner) to make it impermeable. There should always be a check for the presence of land drains – any water infiltrating the ground will flow into land drains and discharge directly to a watercourse. Land drains should be sealed, upslope and downslope, where they cross the site and care taken to ensure the land upslope will not become waterlogged or flooded as a result. For more details on managing land drainage see Chapter 18.

Pond design needs to take into account health and safety provisions such as perimeter fencing, access for emergency vehicles, signage and flotation equipment.

Emptying the pond

Although most permanent balancing ponds (for example on a road scheme) are exempt from the requirement for discharge consents, a consent will be required if the pond is used during construction. This ensures the greater risk of pollution is managed. At the end of construction the consent can be revoked or transferred to the client. See Chapter 13 for details on how and when to apply for consent.

For installation of an outfall, the guidance on outfall design provided in Section 19.8.3 below should be followed. When pumping out water from the lagoon, the pump should be kept off the base of the pond to avoid disturbing sediments. See Chapter 21 for guidance and illustrations on dewatering excavations.

> Contractors must have consent to discharge trade effluent from a settlement pond during site works, even if the pond is part of the permanent works. Only permanent drainage is exempt under the Highways Act 1980.

Figure 19.2
Top left: *unlined excavated pond adjacent to working easement*

Figure 19.3
Above: *series of lagoons (courtesy Young Associates)*

Figure 19.4
Left: *new unlined bunded lagoon*

Maintenance

Outfall structures should always be checked and maintained. Operation of any cut-off valves or penstocks should be tested regularly. Long-term ponds (one year or more) may require management to prevent algal blooms and excessive vegetation growth. Barley straw bales can be placed in the pond to help control algae, while clearing the pond during the summer may prevent excessive build-up of vegetation. It is widely accepted that balancing or settlement ponds can provide wildlife habitats, but this is secondary to fulfilling an engineering purpose. Regular maintenance can help prevent good habitats developing in temporary ponds, which will avoid delays in decommissioning. Nevertheless, on completion of a project care should be taken to ensure that protected species (notably newts, water voles and frogs) or habitats are not harmed when the pond is reinstated.

19.2.3 Settlement tanks

Settlement tanks are purpose-made structures to contain water for the removal of suspended solids. They operate on similar principles to settlement ponds but are portable and smaller. The tanks generally have a series of baffles over which water flows. When installing settlement tanks to remove silt, the baffles should be fixed at the bottom of the tank to ensure the water flows **over** them. Baffles that are open at the base of the tank are designed to allow water to flow **under**, thus trap floating oil and other hydrocarbons. Some tanks are supplied with one of each type of baffle. Settlement tanks can also be set up using skips in series with v-notches for the outlet (see Figure 19.5).

A modified tank is commercially available that incorporates lamellar plates within the tank enabling a water flow rate comparable to a much larger lagoon and sedimentation rates of up 99 per cent, cleaning up to 50 m³ an hour.

Size the pump rate and the tank correctly, otherwise the tanks will not work. Aim for a slow flow over each baffle: if the water is churning in the tank it is not working effectively.

Figure 19.5 *Skips used as settlement tanks (courtesy Hyder)*

A responsible person should appointed to monitor the tank operation and check the outflow is clear on a daily or more frequent basis. Arrangements will need to be made to empty the tank of settled solid materials regularly and dispose of it correctly.

19.2.4 Sizing settlement facilities

Suspended solids will settle out only when the water is still. Usually it is necessary to retain the water in the settlement tank or pond for several hours to allow the suspended solids to settle out. Retention time depends on the particle size, disturbance of the water, depth of water, temperature and particle density. Although it can be calculated (Masters-Williams *et al*, 2001), the particle size and density data are not usually readily available. A rule of thumb indicates that a retention time of 2–3 hours is adequate. However, finer particulate matter may require several days or more and therefore larger settlement facilities. Particles less than 2 mm (eg clays, chalk, coal shale) may never settle out and will require addition of a flocculant (see Section 19.2.6). In ideal conditions retention time is a function of settlement velocity and depth (Masters-Williams *et al*, 2001):

$$\text{Retention time} = \frac{\text{Depth of tank or pond}}{\text{Settlement velocity}}$$

Table 19.2 shows theoretical values of retention time for continuous flow conditions (ie flow in equals flow out). Settlement efficiency can only be increased by increasing the surface area or decreasing the outflow.

Table 19.2 *Theoretical range of retention times for a variety of particle sizes*

Water depth	Retention time (settling velocity)				
	Fine clay (0.001 mm/s)	Fine silt (0.02 mm/s)	Medium silt (0.05 mm/s)	Coarse sand (30 mm/s)	Flocculated silt (10 mm/s)
0.5 m	6 days	7 h	3 h	16 s	50 s
1 m	11 days	14 h	5.5 h	33 s	2 min
2 m	23 days	1 day	11 h	1 min	3 min

A more site-specific idea of the retention time needed can be obtained by leaving a sample of the effluent to stand in a clear jar or bottle and timing how long it takes to clear. Where a consented discharge limit is in force, the sample should be compared to a prepared sample of a concentration equal to the consented discharge.

Because of the time required for suspended solids to settle, several ponds or tanks in series may be needed. A typical example of a multi-pond arrangement is shown in Figure 19.2. See Chapter 18 for guidance on estimating the volume of runoff from a site.

Settlement facilities are best utilised as part of a comprehensive set of control methods. If the suspended solids are particularly fine, or the volume of water very high, no size of lagoon or conventional type of treatment will work. In this case, or if there is no space for alternative suitable treatment, specialist treatment can be used in the form of flocculants or a dynamic separator (see also Section 19.2.6).

19.2.5 Filtration

There area two methods of water filtration – using the flow of water, rather than keeping it still – to remove silts. The first comprises techniques used to trap sediment as water is flowing across site or along channels (see Chapter 18). The second is filtration by pumping water through steel tanks or skips filled with a suitable filter, such as fine single-size aggregate (5–10 mm), geotextile or straw bales. This is an easy and cheap option, suitable for treating discharges with fairly coarse particles, but careful control should exercised over the effluent quality and a mechanism to close down the flow should be incorporated.

An alternative is to construct a simple straw bale "tank", lined with geotextile. Water is pumped into the centre and allowed to filter through the tank walls and thence to flow to ground, a drain or a watercourse (with appropriate consents). This is particularly suitable for temporary "low-risk" discharges adjacent to a watercourse or drain.

19.2.6 Specialist treatment equipment

In certain circumstances, specialist treatment equipment is required. Dedicated commercially available equipment such as centrifuges, dynamic separators or chemical dosing systems are particularly suitable for very fine suspended solids (eg grouts, clay, shale), very large volumes of water or where there is insufficient space for conventional treatment methods. In these circumstances, they are viable treatment methods to include in linear construction sites.

Flocculation

This solution is only recommended when no alternative treatment is available. However, it is very often the optimum solution for treating large volumes of silty water where space is restricted or particle size is too small to settle out, and is therefore highly suitable for use on linear schemes. Very fine particles (typically less than 0.001 mm) will never settle out of liquid without some chemical addition.

> Before work began on a 30 km pipeline, the Scottish Environment Protection Agency (SEPA) actively encouraged the use of flocculants for the treatment of silty water runoff as a result of experience on other linear projects.

Fine particles carry a negative surface charge. A flocculant, in this case, is a positively charged solid, powder or liquid. Flocculating agents increase the rate of settlement of suspended solids by "pulling together" smaller particles into larger, and therefore heavier, particles. Three chemical types of flocculant are available:

- anionic, which have a high affinity for solids, so any remaining concentration in the water is very low
- cationic are recognised as flocculants with greater toxicity implications for fish and other aquatic organisms
- non-ionic types degrade readily in aquatic systems.

Another process, coagulation, is chemically slightly different in that particles do not adhere to each other, but the end result is still settlement of suspended solids. Coagulants, typically aluminium (alum) or aluminium chloride, are non-toxic to aquatic organisms when used correctly.

The design of the system, size and dosing rates etc need careful consideration and supervision to avoid failures of the system. It will also be necessary to gain approval from the relevant environmental regulator before flocculants are employed. The settled sludge produced may be contaminated with metals or other compounds attached to the silt particles and should be disposed of in accordance with waste legislation.

> Always seek specialist advice on the design and use of any flocculant or coagulant system and disposal of the resulting sludge.

Liquid flocculant

Flocculation generally requires two, but sometimes up to four, settlement ponds or tanks. Water is collected in a retention pond (or similar) and a known volume is then pumped to a tank or pond where the flocculant is added. Commonly used flocculants include aluminium sulphate, aluminium chloride and ferric sulphate. Over time the particles come together and settle out. This may take up to 24 hours, depending on the soil type, flocculant used, the dosage, the pH, and the ability to minimise settling tank disturbance. The resulting sediment/soil should be removed regularly with a vacuum truck and may require disposal as contaminated waste. In some circumstances, the sediment may be suitable for use in on-site landscaping or disposal with general excavated site material. It is essential to seek advice from the environmental regulator before disposal or reuse. The following case studies illustrate the use of flocculants with both tanks and ponds, together with differing methods of dose calculation.

Disadvantages of liquid flocculant include:

- high concentration supplied to site requires dilution before use, requiring specialist mixing equipment and a power source
- after pre-dilution, liquid has a limited shelf-life
- complete mixing is still difficult because of high viscosity (Beca, 2004)
- required intensive management to monitor and supervise the system.

Case study 1. Use of flocculant on pipeline scheme – mechanical dosing

The solution chosen to treat large volumes of silty runoff was construction of a water collection, treatment and disposal system. This involved excavation of two unlined collection lagoons, gravity- and pump-fed balancing lagoons, a flocculant dosing system and two settlement lagoons. This comprehensive system was designed to prevent pollution of a watercourse feeding a potable supply reservoir.

Lagoon alongside pipeline easement

The water management system served a 1 km length of pipeline on a gradual slope from moorland down to a road crossing. Runoff was diverted to the sides of the spread and collected in a channel. This was piped over a small ditch up to the point where it left the spread and was channelled into the balancing lagoon. Runoff below this point was collected in two lagoons and pumped up to the balancing lagoon. The outflow from the balancing lagoon was regulated through a penstock valve. The treatment chemical (flocculant), in an intermediate bulk container (IBC), was stored with the pump in a lockable steel cabin. The dosing pump delivered a measured flow of flocculant into the treatment channel and was manually linked to the penstock setting. The treatment channel was produced from polythene membrane sheeting (Visqueen), formed to create as much agitation as possible to mix the flocculant into the water. The treated water was then directed into the first of two large settlement lagoons in series. The discharge was made from the second settlement lagoon to a watercourse.

Flocculant dosing pump and IBC inside steel cabin

Lagoon and penstock valve and dosing line in foreground

The system worked very well until a few days of particularly heavy rain. This caused a minor landslip resulting in the dosing line becoming detached. As a consequence, a large quantity of untreated water moved through the system and polluted the watercourse.

This system required a significant resource to manage and maintain it. However, without it, the company would have faced almost certain legal action.

Left: *silty water in the right-hand lagoon, clear treated water in left-hand lagoon*

Solid flocculant

Solid flocculants, usually in the form of "floc blocks", are placed in a channel of flowing water and dissolve in the flow, providing a dose of the chemical. The performance of the blocks is dependent on the flow and the suspended sediment concentration.

Disadvantages include:

- high bedload of sediment can accumulate against and even bury the blocks
- flow rates may need to be restricted to achieve adequate dosing
- hard to set up a system that provides optimal dosing for variable stormwater flow.

Advantages include:

- high potential for sediment removal in regulated flows of consistent water quality (eg small catchments)
- requires less time for sediment to settle than liquid coagulant, so retention tanks or ponds can be smaller.

Case study 2. Use of flocculant on pipeline scheme – jar test

On a shorter length of the pipeline mentioned in Case study 1 the solution was to collect the silty water at the side of the working area, pipe it over stream crossings in a twin-wall pipe, and collect it in a large (approximately 50 m × 20 m × 3 m) unlined storage lagoon.

At intervals the water was pumped down into an unlined treatment/settlement lagoon (approximately 25 m × 25 m × 3 m). When this was full, the contents were stirred using a pump strategically placed to create a circulation within the lagoon.

Once mixing had been achieved, a 2 litre sample of the water was taken in a clear bottle. Measured amounts of flocculant chemical were added to the sample until the correct dosage was achieved. Working on the results of this "jar test", the volume of flocculant chemical needed to treat the whole of the lagoon was calculated.

Water sample 15 seconds after adding the flocculant

The flocculant was added to the lagoon at the point where the circulating pump discharged back into the lagoon to enable maximum mixing. Once all the flocculant had been added, the circulating pump continued to mix for several hours. The pump was then switched off and the lagoon contents left to settle overnight.

The next morning, the clarity of the water was checked and, if satisfactory, pumped carefully out of the lagoon into the receiving watercourse.

As with other flocculant dosing systems, a resource is needed both to manage and to maintain it.

Water sample 5 minutes after adding the flocculant

Case study 3. Use of flocculant (after Jurries, 1999)

A site opted to try flocculation when all other measures to control silt and turbidity in site runoff failed to reach the discharge consent requirements. Two pumps in parallel picked up the stormwater runoff and delivered it to the first tank for the addition and mixing of the flocculant.

The prescribed volume of flocculant was added and mixed for a set time in the 64 000 litre tank. The system was controlled by input volume and had the capacity to mix and treat approximately 2000 litres per minute.

Pumps were used to evacuate the mixture to the settling tanks. The extraction system was designed to exceed the input capability. Two baffled 64 000 litre settlement tanks were used for settling. Utilising a separate settling tank allows smaller controlled volumes to be mixed and then combined in a larger settling tank. The solids settled out in the tanks and clean water was discharged by gravity from an outlet located approximately 500 mm up from the base of the settlement tank. The discharge was piped into the adjoining stream, where it was sampled.

Separators

Purpose-made equipment can separate various types of pollutant, including oils and solids, from site water. The manufacturers of these devices produce specific designs based on the peak flows, sediment grading and removal efficiency. Dynamic separators typically retain particles as small as 75 mm and have an overall removal efficiency of around 90 per cent of particles greater than 150 mm. In some cases it may be possible to include a permanent drainage scheme requiring a dynamic separator in the temporary works to reduce cost.

Case study – water treatment plant (source: Hyder)

Dewatering a tunnel construction produced a large volume of water polluted with coarse particles from surface water, fine black shale from drilling and high pH from the grout used. A dedicated treatment plant was installed on site, which combined a silo of sand to filter out coarser particles and a centrifuge to dewater the slurry. Oil-absorbent booms were floated on the surface of an interim tank to remove any hydrocarbons. The final treatment was a controlled dose of sulphuric acid to lower the pH to an acceptable level prior to discharge to the surface water sewer. The sand filter material required regular replacement, but both this material and the shale filter cakes could be simply disposed of with excavated material leaving site. The pH was continually measured and recorded electronically. Regular visual inspections and sampling of the discharge ensured that the equipment was satisfactorily removing the suspended solids in compliance with the discharge consent.

19.3 CONCRETE AND CEMENTITIOUS MATERIAL

The aim should be to prevent cementitious material from entering water on site (and always to prevent it entering watercourses), but there will inevitably be some polluted water generated from concrete washout, batching operations or grouting below ground (tunnelling etc). Chapter 22 contains guidance on minimising pollution from concrete and grouting operations. In general, the treatment options will depend on the volume of water and the availability and treatment of other water on site. The prioritised list below can be used to determine the solution.

1. Consider whether other treatment methods are being implemented on site that will remove suspended solids. If not, a suitable method should be employed to settle out the concrete solids. These can range from settling out in a skip to the use of specialist equipment (all discussed in Section 19.2). If other methods are already in place, point 2 should be considered.

2. The volume of water and length of operation need to be determined by answering the following questions:
   - will large water volumes be generated or are there plans for long-term operations that will produce large amounts of settled-out concrete? If so, water should be treated separately from other site water (or returned to the on-site batching plant if there is one). Large volumes of concrete-polluted water may have to be pumped out frequently or be broken out when dry, so separate treatment will avoid the need for extra maintenance of the existing facilities.
   - if a smaller volume is anticipated, can it be diluted by other water in the existing treatment facilities to a suitable pH level to be discharged? If so, treat the concrete polluted water with other site water. If not, treat it separately (or return to onsite batching plant if there is one).

3. Consider whether any chemical flocculants are being used in existing facilities that may be affected by the introduction of concrete pollution. If so, the concrete-polluted water should be treated separately.

Water that is to be reused in subsequent concrete batching should be chemically suitable (eg not naturally sulphate-rich) to do so. Further details on providing and controlling concrete washout areas is provided in Chapter 22.

It may be necessary, if large volumes of water require treatment, to dose the settled water with acid (powder or solution) to reduce the pH sufficiently for disposal. Dedicated equipment is widely available from specialist water treatment companies. It may be possible to select either an automated dosing system for continuous use or a system requiring manual pumping and dosing for occasional use. In all cases, the facilities should be managed by appropriately trained, dedicated staff, while the possibility of equipment failure should be covered by suitable contingency measures.

19.4 FUEL AND OIL

The aim should be to prevent fuel and oil from entering any water on site (and they must never be allowed to enter watercourses), but there will inevitably be some oil-polluted water generated in site runoff, particularly from fuel bunds, drip trays and refuelling areas. Treatment measures generally take three forms.

1. Provision of **oil separators** to remove hydrocarbons from high-risk areas of runoff.

2. Use of **oil-absorbent materials** to remove small quantities of oil or provide an emergency measure in the event of a fuel spill.

3. Use of **SUDS** techniques to filter and biodegrade hydrocarbons.

Oil also sticks to fine sediments, which will settle out and may cause further ecological harm. Ensure sediments are also removed.

On a linear scheme it is likely that hydrocarbon removal will be required at more than one location and usually where space is restricted, for example at several discharge points, or when working in or near water. Unless the oil removal is required in a long-term facility, portable measures will be more useful.

Oil separators

Oil separation works by the simple principle that oil floats on water. Any structure or device that impedes flow on the water surface can effectively "trap" the oil. A separator may be installed in front of the inlet to a settlement pond, for example, although generally they are installed as part of the permanent drainage works, either at the inlet to a retention pond or in the drainage from roads, car parks and railways.

Portable oil-separation tanks (see "Settlement tanks") have a series of baffles designed to allow water to flow **under** thus trap floating oil and other hydrocarbons. Oil-separating tanks must not be confused with settlement tanks.

A simple temporary measure can be constructed from planks suspended across a narrow ditch, watercourse or drainage channel such that they extend below the water surface to trap any floating hydrocarbons.

Two types of "permanent" commercial separators constructed from prefabricated glass reinforced plastic (GRP) are often used:

- bypass separators are designed for the control of spillages and have the capacity to treat flows generated by the majority of storms (typical rainfall intensity of up to 5 mm/h). Flows greater than this bypass the separator
- full retention separators are specified to treat the peak design flow for a particular system and so are both more efficient and more costly than bypass separators.

Further details on the specification and use of oil separators can be found in PPG3 (EA, SEPA and EHS, 2004h), which is available from the environmental regulators' websites.

Oil absorbents

Oil-absorbent pads are floated on the surface of water to absorb any hydrocarbons accumulated in settlement ponds and tanks as well as fuel bunds and drip trays. Oil-absorbent booms are fixed across drainage channels, settlement ponds and outfalls (Figure 19.6). Booms need to be fixed securely at each end to prevent water flowing around the ends when the water level fluctuates. Absorbent materials should be replaced when they become discoloured.

A recent development is a cotton-based oil sponge impregnated with naturally occurring bacteria that break down hydrocarbons.

The environmental regulators discourage the use of oil dispersants because they increase the bioavailability of the hydrocarbons and may be harmful to the environment.

SUDS

Sustainable drainage systems (SUDS) such as swales and permeable paving can remove hydrocarbons in two ways: first, by trapping fine sediments to which oil adheres; second, by encouraging the natural biodegradation of hydrocarbons. Further details can be found at <www.ciria.org.uk/suds>.

Figure 19.6 *Settlement pond with straw bale filter and oil boom (courtesy Hyder). Note that fluctuation of the water level has lifted the ends of the boom above the water surface, which might allow oil to flow underneath*

19.5 METALS

Heavy metals (lead, iron, zinc etc) may be present in the ground, particularly at contaminated sites and may also be present in groundwater or mine drainage. In certain parts of the country, notably Cornwall, they occur naturally. Other metals, such as aluminium, may be generated by the use of flocculants (see Section 19.2.6).

Metals can adhere to sediment particles, allowing them to be removed by settlement or similar sediment-removal techniques (see Section 23.2). SUDS can also be used to remove 30–40 per cent of metals from water (van Bohemen and Janssen van de Laak, 2003).

Where metals are dissolved in the water, traditional filtration through sand filters, aerating (see case study) or dosing with a coagulant or precipitant (see Section 19.2.6) can be effective. The latter typically causes a pH change (to alkaline), which precipitates out certain metals from solution. One disadvantage is that this process can also mobilise other metals. Specialist advice and approval should be sought from the environmental regulator before coagulants or precipitants are used.

In all cases, the resulting sludges may be contaminated, so advice should be sought from the environmental regulator before disposal.

> **Contaminated minewater on pipeline scheme** (source: Alfred McAlpine)
>
> During construction of a pipeline through an area of mineworkings, the contractor encountered a significant ingress of iron-contaminated groundwater into a deep-thrust pit for an augur bore crossing. Two large unlined lagoons (approximately 60 m × 25 m × 5 m) were constructed, involving extra land take, to hold and treat the ochreous water. The water was pumped from the thrust pit into the lagoon with as much agitation as possible. This was done by discharging it from several metres above the surface. By this method, air was introduced into the water, which oxidised the dissolved iron to produce iron oxide. This settled out as an orange sediment. The clear water was then discharged slowly into a nearby ditch.
>
> The approach was wholly successful and allowed the crossing to be made without causing pollution of the watercourse.

Resultant levels of metals would still need to be compared to those allowed in the discharge consent. Where limits are stated only for oils and suspended solids, the environmental regulator should be aware if there are other contaminants present.

19.6 AMMONIA AND OXYGEN LEVELS

High BOD, COD and ammonia levels are associated with a lack of oxygen in the water, which affects its ability to support fish and other water fauna. Where water quality is poor because of lack of oxygen, for example as a result of dewatering near a landfill or river deposits or even die-back of vegetation in ponds, the water should be aerated. This can be simply achieved by allowing the water to flow over a series of weirs, pumping air through a perforated underwater hose or using commercial bubble diffusers, compressors and aerators.

19.7 SEWAGE

Under no circumstances should sewage be discharged to the ground or to a surface water drain or watercourse without prior treatment and consent from the environmental regulator. The preferred method of disposal is to the foul sewer (see Section 19.8). Where no mains drainage is available, suitable alternatives are septic tanks or a package sewage treatment plant. A cesspool to contain the sewage for removal and treatment off site is the least favoured option (and is not permitted in Scotland). Further guidance is available in PPG4 *Disposal of sewage* where no mains drainage is available (EA, SEPA and EHS, 2004a).

Biofiltration treatment methods use naturally occurring or introduced vegetation such as reeds (phragmites) to break down and filter sewage-polluted water. The effectiveness of biofiltration is seasonally dependent and not suitable for treatment of non-biodegradable pollutants. Suitable vegetation can be incorporated in filter strips, swales and shallow wet ponds, but careful advanced planning and consideration is required before a decision is taken to implement this system.

19.8 DISPOSAL OPTIONS AND TEMPORARY OUTFALLS

Water should not be discharged without prior permission and the discharge must comply with any conditions specified in that permission. Linear construction projects are most likely to require multiple discharge points, which may involve more than one sewerage undertaker or environmental regulator. Guidance on when and how to apply for consents and authorisation is provided in Chapter 13.

> In most circumstances consents take months to obtain, so plan ahead to avoid delays.

In deciding how and where to dispose of water from a site, the scheme should be considered as a whole, the best practice hierarchy (below) followed and the checklist of practical considerations (see opposite page) reviewed.

1 **Infiltration**
  The preferred option for sustainable water disposal is to infiltrate as near to source as possible; see Building Regulations 2000 and Planning Policy Guidance Note 25 (DTLR, 2001). Guidance on infiltration techniques is provided in Section 19.2.1.

2 **Discharge to surface water**

If infiltration cannot be achieved, the next best option would be to obtain consent(s) to discharge to a local watercourse or surface water sewer, subject to appropriate treatment.

3 **Disposal to foul sewer**

If disposal to a watercourse is not feasible or acceptable because of water quality issues, agreement to connect to a foul sewer should be obtained.

4 **Disposal as waste**

Water that is too polluted for the sewerage undertakers to accept and that cannot be treated on site must be stored on site as controlled waste in clearly labelled tanks before it is pumped into a tanker for off-site disposal, which may be costly. Wastes must be handled and disposed of according to duty of care legislation.

Considerations for choosing disposal methods and locations

1 The volumes of water involved, based on the size and characteristics of the site and catchment.
2 Site topography (to determine where water will collect, and whether pumping will be required).
3 Inclusion of features such as balancing ponds in permanent design.
4 Whether space is available for temporary storage and treatment.
5 Location of suitable receiving water.
6 Location of foul sewer connection.
7 Degree and type of any pollution.
8 Agreement from the environmental regulator or sewerage undertaker (formal permission will be required).
9 The charges to be levied by the regulator and sewerage undertaker.

Case study - seven consents needed for road project (source: Hyder)

A 14 km road project was constructed through the centre of a major city. The project included a 7 km-long bored tunnel and a river bridge. Seven discharge consents had to be obtained for the project – each selected for a particular quality of water at particular locations – one via settlement lagoons into the river, one for clean dewatered groundwater direct to the river, one via a water treatment plant to surface water sewer, two to surface water road drains and two to foul sewers.

19.8.1   Discharge to surface water

Discharges to surface waters require written permission in the form of a discharge consent from the environmental regulator. Details on how and when to obtain a consent and the likely conditions that may be attached to it are provided in Chapter 13. The environmental regulator levies a charge depending on the volume of the discharge.

> **Case study - invertebrates wiped out by silt pollution** (source: *The ENDS Report*, Jan 2005)
>
> A civil engineering company had to pay a fine of £8000 and the Environment Agency's costs of £1185 for allowing silt to smother the bed of a tributary of the River Loddon, in Reading, killing the stream's invertebrate life. The company had been employed to lay down sewers for a new housing development. During the dewatering stage of laying down the foul and storm sewer pipes, water that had not been sufficiently de-silted was pumped directly into the river. The environment officers advised them that they had committed an offence.
>
> Returning to the site, the environment officers traced where pipes had been laid down towards the stream and saw a large amount of silt accumulated in the area. The company did not have consent from the Environment Agency to discharge to the stream at this location. The company offered to clean up the silt and removed the pipes.

> Ensure the discharge complies with all conditions given in the discharge consent.

19.8.2 Outfalls

Discharging water at high, or even moderate, velocities into a watercourse can cause disturbance and erosion of the banks or bed, particularly if the discharge is sporadic. This results in the water being contaminated by suspended sediment. The exit velocity at the outfall should be reduced using baffles, blocks placed in the outfall apron or an energy-dissipater (see Figure 19.7). The same precautions should be taken when overpumping water along a watercourse (see Section 20.7 and Figure 20.12).

Continuous scouring action can cause long-term damage to watercourses. Precautions should be taken to avoid such damage occurring at the outfall, as well as on opposite banks, using geotextile, stone or plastic sheet (see Figure 19.8).

Figure 19.7 Above: *baffles on discharge hoses (courtesy Transco)*

Figure 19.8 Right: *bank protection for discharge outfall (courtesy Transco)*

Much of the industry guidance available on best practice design for permanent outfalls to watercourses, including that recommended by the Highways Agency, Environment Agency/SEPA, Rivers Agency, British Waterways and water companies, can be applied to temporary (particularly long-term) outfalls. Where a settlement pond is installed, the outfall should be placed as far away from inlets as possible.

Outfalls should be angled at 45° to the water flow; small pipes (less than 300 mm diameter) can be at a maximum of 90° to the flow. It may be necessary to install a flow restrictor and/or flow meter to limit and/or monitor the flow rate of gravity systems to that allowed in the discharge consent. There should be some method to close or isolate the outfall in the event of a pollution incident.

Safe access to allow sampling and monitoring of the discharge should be provided. This could either be at the outfall itself or by means of a manhole.

19.8.3 Disposal to foul sewer

Pumping to foul sewer poses the least risk of environmental pollution but is not always cost-effective nor is it sustainable for conservation of local water resources. The distance to a connection at rural sites may also be a limiting factor and necessitate alternative disposal options.

Discharges to public foul sewer require written permission from the statutory sewerage undertaker in the form of a trade effluent consent; alternatively, the contractor may enter into a trade effluent agreement with the sewerage undertaker. A separate consent or agreement may be needed for each discharge along the site. Chapter 13 gives details on how and when to obtain a consent and the likely conditions that may be attached to it.

It is an offence to discharge to the foul sewer without consent, or to discharge volumes or substances other than those agreed. This is because the typical biological sewage treatment is specific to the normal content of the wastewater and may not remove oil, dissolved metals or chemicals. Any effluents discharged that damage the sewers, generate a health hazard for sewerage workers or the general public, or inhibit or kill the micro-organisms that carry out the sewage treatment process can be traced to their source and the responsible parties held liable.

> Ensure you have a trade effluent consent/agreement in place and the discharge complies with all conditions in it.

In general, sewerage undertakers prefer to receive small volumes of poor-quality effluent rather than large volumes of unpolluted effluent (they are unlikely to agree to accept large volumes of dewatered groundwater, for example). This is often the opposite for discharges to controlled waters. If only small quantities of wastewater are generated, it should be pumped into clearly labelled tanks for removal at a later date by a specialist contractor in accordance with duty of care legislation.

It is essential to identify and connect to the foul sewer, and not the surface water sewer.

> **Companies fined for river pollution from sewer system** (source: *The ENDS Report*, Mar 2005)
>
> A construction company was fined £4000 after polluting a Welsh river with sediment. A civil engineering company was also fined for its part in the incident. An Environment Agency officer was out on a routine inspection of the River Clydach, when he saw that the river was heavily discoloured. Tracing the pollution to a sewer overflow, he discovered the pollution was coming from excavation work in the town centre. The work related to a new mains sewer system. The two companies had accidentally connected their wastewater pipe to a surface water drain. Both companies pleaded guilty to causing polluting matter to enter the river and were each fined £4000 with £650 costs.

The sewerage undertaker levies a charge appropriate to the level of service provided. Charges for trade effluent are based on the Mogden formula, which is based on the characteristics (volume and strength) of a customer's discharges. These, in turn, determine the level of treatment needed and therefore the costs involved. Further information on charging can be found at <www.ofwat.gov.uk>.

The volume of flow is the most important factor in determining the bill. More often than not the volume of effluent is assumed to be the same as supply, rather than being monitored. If any of the water supplied to site is not discharged to the sewer, it is still paid for as such. If a large volume of water is supplied to the site or discharged to several sewers, the bill can be high.

> **Checklist – water disposal**
>
> Financial benefits can be gained in reducing sewage bills by:
>
> - reducing water supply
> - reducing volume of effluent discharged
> - monitoring flow rate/volume of effluent.

> **Checklist – water disposal**
>
> 1. Review the site characteristics (topography, watercourses, foul sewers) and construction activities to decide on the number and location of effluent discharge points needed along the whole route.
> 2. Consider the priorities:
>   - infiltration
>   - disposal to a watercourse
>   - disposal to sewer.
> 3. Apply for consents from the sewerage undertaker and environmental regulator for each discharge point see Chapter 13).
> 4. Manage the supply and use of water on site to reduce the volume requiring disposal.
> 5. Minimise erosion and scour by providing an adequate, accessible outfall.
> 6. Set up measures to monitor the discharge (see Chapter 14).

20 Works in or near water

Construction works that take place in or near water differ widely in complexity, scale and timeframe. Similarly, the water environment itself ranges from ditches to major rivers. Examples of working in or near water include open-cut crossings, bridge piers, culverts, haul route crossings, headwalls, flood defences, river diversions, cofferdams, piling, directional drilling, bank-side maintenance, bridge cleaning and repairs.

This chapter describes what should be considered before starting work, best practice methodologies for common activities and pollution prevention measures. It should be read in conjunction with the guidance set out elsewhere in this book. Works involving groundwater management, ie below the water table, are discussed in Chapter 21.

Guidance on the use of specialist plant and techniques for in-river working, such as jack-up barges, floating cranes and pontoons, is provided in CIRIA C584 *Coastal and marine environmental site guide* (Budd *et al*, 2003).

20.1 PLANNING THE WORKS – LEGAL REQUIREMENTS

Written permission from the environmental regulator is required to carry out work in or near watercourses in England, Wales and Scotland (in Northern Ireland the Rivers Agency should be consulted). Section 13.4 gives full details on the consents required for works in or near water. It is important to check the contract (and any environmental statement) for specific requirements relating to the works. Construction techniques and best practice should be discussed with the environmental regulator and key subcontractors.

> Before planning the method of work, contact the environmental regulator and ensure all required permissions are in place. You may need to allow up to **four months** in the construction programme for some consents to be granted.

Separate consents may be required for dewatering, overpumping or discharging as well as for working in or near water. (In Scotland, however, one "authorisation" will apply to all the proposed activities.) Consents may refer to any of the following:

- flood defence consent
- land drainage consent
- discharge consent
- transfer licence
- authorisation.

If British Waterways owns or manages the waterway (eg a canal) it must be consulted on any proposals. Its *Code of practice for works affecting British Waterways* (British Waterways, 2005) should be followed; a formal agreement may be required to allow works to proceed.

If an internal drainage board (in England and Wales) manages the waterway it must be consulted. The Environment Agency and also the Association of Drainage Authorities (<www.ada.org.uk>) will be able to confirm whether the IDB manages watercourses in the area.

> **Key guidance**
>
> See Chapter 13 for further information on why, when and how to apply for consents.
>
> British Waterways (2005). *Code of practice for works affecting British Waterways*, <www.britishwaterways.co.uk/images/COP_2005.pdf>
>
> DTLR (2001). Planning Policy Guidance Note 25, *Development and flood risk*, to be replaced in 2006 by a new planning policy statement, <www.odpm.gov.uk>
>
> EA, SEPA and EHS (2000). PPG5 *Works in, near or liable to affect water*
>
> Scottish Executive (2000). *River crossings and migratory fish: design guidance*, <www.scottishexecutive.gov.uk/consultations/transport/rcmf-00.asp>

Once consents are in place the attached conditions should be checked. At sensitive or protected sites the conservation agency should be consulted and a method of work agreed that protects the ecological interest of the watercourse.

> **Potential restrictions on works in or near water**
>
> Consent is likely to be required for temporary or permanent works in or near a watercourse or on a floodplain.
>
> Certain construction methodologies may be required in order to safeguard the integrity of canals, designated ecological sites or amenity interests.
>
> A Food and Environmental Protection Act (FEPA) licence will be required for works below the mean high water mark (eg crossing a tidal river).
>
> Works may not be permitted during the fish spawning season, which runs from around mid-October to mid-April.
>
> Fish rescue may be required immediately prior to the works.
>
> Works may not be permitted within a certain distance upriver from the sea during the summer because of possible adverse impacts on coastal and bathing water quality.

> **Case study – flood defence consent** (source: Hyder)
>
> A replacement sewerage pipeline, 30 km long, was constructed in a rural area. The route crossed numerous field boundaries, many of which were defined by a stream or ditch, and three main rivers. In consultation with the local Environment Agency office, costs and time were saved by developing a generic method statement for each method of crossing and applying for just three land drainage consents (now flood defence consents) covering 15 watercourse crossings along the pipeline route.
>
> 1. Temporary culverted haul route access across streams.
> 2. Open-cut stream and ditch crossings.
> 3. Open-cut river crossings, with flume to carry water flow.

20.2 POLLUTION CONTROLS

Works carried out adjacent to or in water pose a significantly greater risk of water pollution than inland works. They should be subjected to a risk assessment and a detailed method statement prepared (one will be required for consent applications) that includes pollution prevention measures. All subcontractors and staff need to be made aware of its contents. In addition to the guidance throughout this chapter, which includes specific guidance on construction methodologies, the following pollution control measures should be adopted when working in or near water (see Sections 20.2.1–20.2.5).

20.2.1 Emergency plan

An emergency plan for the whole site should be prepared, containing particular contingencies for a pollution incident in or adjacent to water. Emergency equipment such as absorbent mats, booms, straw bales, geotextile matting and rags should be kept adjacent to the watercourse (rather than in the site cabins), and on both sides if the work involves crossing the watercourse. It is also important to identify flood-prone areas, obtain weather forecasts, register for automatic flood warnings where appropriate and maintain contact details of any downstream users, such as fisheries. Detailed guidance on all these issues is provided in Chapter 15.

20.2.2 Silt generation

Silt can be generated by disturbance of the bed material, excavation of the banks or runoff from the working area, haul routes and crossings (see also Sections 20.3 and 20.9).

Straw bales are effective at filtering out coarse particles from flowing water (Figure 20.1). They can be fixed with stakes or weighed down in deeper water to prevent them floating away. Straw only lasts for a short time and eventually biodegrades. When the bales become silted up they should be replaced, taking care to avoid remobilising sediments (see case study below). The disposal of these silted-up bales should be agreed with the local Environment Agency office before work takes place.

Case study – bank protection works cause silt generation (source: Nuttall)

The banks of a brook had slipped and required stabilisation at the toe of the slope. The works had the potential to generate silt in the brook. The solution was to place straw bales across the brook at five places downstream, creating a series of settlement lagoons. This was effective at containing the silt generated by the works. However, removal of the straw bales tended to remobilise the sediment. It was essential to remove the upstream rows of bales first.

Figure 20.1
Straw bales filtering silt in a newly constructed stream (courtesy Hyder)

Effective trapping of sediments on the river bed can be achieved using silt mats. These are flat pads made of layers of biodegradable materials that are fixed to the river bed downstream of the works (Figure 20.2). The mats trap sediment being transported along the river bed. When construction is complete the mats are designed to be rolled up and removed, and can be seeded and used as bank stabilisation without the need for disposal. Removal may cause some silt leakage to occur, which can be controlled by temporarily installing mats downstream as the laden mats are being taken out. Mats should be installed across the full width of the watercourse in accordance with the manufacturer's guidelines.

Figure 20.2 *Silt mat (courtesy Mowlem)*

> **Case study – inadequate silt filtration measures** (source: *The ENDS Report*, Jun 2001)
>
> A pipeline company and a construction firm were each fined for their parts in separate water pollution incidents. The pipeline company was contracted to lay a pipe under the river Kennet, near Mildenhall, Wiltshire. An investigation found that silt and chalk from construction activities had entered the river, turning it white for 1.5 km. Attempts by the company to filter out the particles using straw bales had proved futile. The pipeline company pleaded guilty to causing polluting matter to enter controlled waters. The pollution occurred only 200 m upstream of a trout farm and caused distress to many fish. The Kennet is a chalk stream and site of special scientific interest. The company was fined £6000 and had to pay £2000 costs.

20.2.3   Fuel and oil

A designated area or areas for **fuel** storage and refuelling should be set up according to best practice at least 10 m from any watercourse to minimise risks from leaks and spills (the guidance in Chapter 16 should be followed). An absorbent containment boom should installed across the watercourse or around the works, securely and closely anchored to the banks or working platform.

Commercially available oil-absorbent booms should be installed across the watercourse to contain any fuel or oil spill. Booms can be attached in sequence to cross wide watercourses. The boom needs to be secured to the bank, with ends turned upstream to prevent any oil escaping around the ends (see Figure 20.3).

Refuelling over water

The refuelling of plant and equipment should, wherever possible, be carried out on land to minimise the risk of water pollution. However, this is not always practical, particularly for waterborne craft, floating plant and equipment and works taking place from jetties etc.

All refuelling, whether on land or over water, should follow the best practice guidance set out in Chapter 16. Refuelling over water requires additional care. A delivery pipe with screw-fixed coupling and an automatic shut-off valve should be used. Sufficient time should be allowed for fuel to drain down before this connection is removed. A drip tray and absorbent pads should be in place at the point of refuelling to deal with any spills and drips. Booms and emergency spill kits should be made available, with staff trained to competency in their use (see Chapter 15). Refuelling and emergency arrangements should be clearly displayed and communicated to the workforce and tested periodically for effectiveness.

20.2.4 Concrete

Chapter 22 provides guidance on the use of concrete, including the use of pre-cast structures and quick-setting mixes, careful controls on the transport and placement of wet concrete and underwater concreting. Specific information on the control of bentonite during trenchless construction is provided in Section 20.4 below.

Figure 20.3   *Correct installation of oil boom*

20.2.5 Monitoring

Monitoring water quality before, during and after works is an important part of the risk management process. Even where monitoring is not required as part of contract documentation or consents, it is always worthwhile carrying out baseline monitoring of the watercourse before construction works begin. See Chapter 14 for further information on monitoring.

20.3 ACCESS AND HAUL ROUTES ACROSS WATER

Entry into water should be avoided wherever possible. If plant or vehicles have to make repeated crossings of a watercourse, a temporary haul road bridge or flumed/culverted crossing can be erected. For narrow streams and ditches, culverted crossings are most appropriate (see Figure 20.4 and case study below). For larger watercourses, or where heavy plant is required to use the haul roads, more robust crossings should be established (see Figure 20.5).

> **Case study – haul route stream crossing site procedure** (source: Hyder)
>
> Where the pipeline route crosses streams these will be temporarily culverted to obtain access. At watercourse crossings the working width will be reduced to approximately 8 m to minimise the area disturbed. Prior to carrying out any works close to watercourses, a flood defence consent will be obtained from the Environment Agency (EA). To establish the crossing points on narrow streams, a culvert pipe will be laid to carry the water, and at either end the culvert will be surrounded by a clay bund. The culvert pipe will be laid at winter water level to ensure that water flow is unimpeded along the stream. Straw bales will also be positioned at either end to prevent suspended solids moving along the watercourse. Water between the clay bunds will be pumped out on to the land if necessary. Water quality monitoring for suspended solids, pH and oils will be undertaken before, during and after stream crossings to determine if any adverse impacts are occurring. Measures to reduce or eliminate any identified adverse impacts will be discussed with the EA, Countryside Council for Wales (CCW) and internal drainage board (IDB) and implemented where appropriate.
>
> The culverted area will then be backfilled with inert (preferably site-won) material, and wooden "bog mats" laid to enable plant to drive over the watercourse.
>
> On completion of the whole works, the temporary culverts will be removed from the watercourse, including the clay and backfill. The watercourse will be reprofiled using an excavator blade to seal and puddle the clay providing an impermeable layer and the banks reinstated according to the specifications and requirements of the EA, the IDB and CCW. The bank will then be left to revegetate naturally. The straw bales will remain in place until the watercourse has been reinstated to prevent the movement of suspended solids along the watercourse.

Figure 20.4 *Culverted haul route crossing (note straw bales and geotextile to filter silty runoff) (courtesy Morgan Est)*

> Always ensure that culverts are sized to take the anticipated maximum flood flow in a watercourse (Figure 20.5). Follow the guidance in Section 18.3 to determine the flood flow.
>
> Locate the culvert pipes on the bed of the watercourse to avoid restricting low water flows and to allow the free passage of fish.

Figure 20.5 *Haul route culverts sized for maximum flood flow (courtesy Costain)*

Where an existing bridge structure is used as a haul road, mud and debris should not be allowed to build up. Straw bales or sand bags should be placed along the sides (see Figure 20.6) to prevent silty water running off into the water below.

Figure 20.6 *Sand bags along existing bridge protecting river below from haul road runoff (courtesy Morrison Construction)*

20.4 TRENCHLESS CONSTRUCTION

Where it is necessary for pipelines or cables to cross watercourses, the regulators are increasingly finding open-cut excavations unacceptable. In most cases, the use of trenchless techniques in the form of directional drilling is presumed (British Waterways, 2005). The equipment technology has advanced considerably, leading to increased confidence in the techniques. Benefits of horizontal directional drilling are minimal impacts on the watercourse, reduced levels of reinstatement, faster installation, production of significantly less waste spoil and fewer seasonal restrictions.

The major applications for directional drilling include new sewerage, sewer relining, gas and water mains, oil pipelines, electricity and telecommunications cables and culverts, in particular when crossing rivers, streams and canals. Systems allow installation of pipes in the range of 150–3000 mm diameter. The technique for pipelines up to 900 mm diameter constructed in this manner is normally referred to as microtunnelling and above this diameter as pipejacking, but the principle remains the same (Pipe Jacking Association, 1995a).

Directional drilling is a steerable method of installing pipes in a shallow arc using a surface-launched drilling rig. To install a pipeline, thrust (drive) and reception pits are constructed, usually at manhole positions either side of the watercourse. Drive and reception pits should be as far from the water's edge as possible – a distance of at least 5 m is required by British Waterways (British Waterways, 2005) and up to 10 m by environmental regulators, depending on the watercourse and flood hazard.

A pilot bore is drilled, using bentonite or another form of liquid grouting medium to lubricate the drill or support the hole during boring. The pipe is strung and welded in a straight line at the reception pit. The hole is enlarged to the required size by a back reamer pulled back through the bore into the entry pit, installing the product pipe at the same time.

Bentonite and grout can cause serious harm to the water environment, as they are highly alkaline. Use of these materials must be carefully controlled to avoid breakout of bentonite in the river bed (see case study below) or spillage and runoff from tanks and plant at the drive shaft. Bentonite should be recycled throughout the process, but must be disposed of as controlled waste at the end of construction. (See also Chapter 22 for guidance on concrete and grouting operations.)

> **A detailed site investigation is essential. The more comprehensive the results from the site investigation, the greater the chance of significantly reducing the risk of a bentonite breakout.**

Money and time spent on a site investigation is never wasted, and it should involve the directional drilling contractors, who are experienced in identifying the best borehole locations. There will always be an element of risk, as a surface breakout of bentonite is dependent on the configuration of cracks in the bedrock and the degree of pressure exerted during drilling activities, so a detailed contingency plan is essential. One specialist drilling contractor estimates that a breakout occurs in one in 10 jobs (Pipeline & Plant Construction Group Environmental Forum, 2003).

> **Key guidance**
>
> Pipe Jacking Association (1995b). *Guide to best practice for the installation of pipe jacks and microtunnels*
>
> British Waterways (2005). *Code of practice for works affecting British Waterways*, Sections 6.3 and 6.4

Case study – bentonite breakout from directional drill

For construction of a 30 km gas pipeline, the contractors decided to use a horizontal directional drill (HDD) technique to cross a main road, a stream, a canal and a motorway. This meant drilling to accommodate a gas pipe diameter of 1050 mm for a horizontal distance of just under 1 km. All went smoothly until final reaming, when the grout lubricating the operation broke out to the surface and emerged at the side of the stream, causing discoloration (Figure 20.7). The drill company halted the operation, immediately, which stopped the pollution, and called the Environment Agency.

Figure 20.7 *First grout breakout controlled with sand bags*

After discussion, the parties agreed to carry on drilling while keeping a permanent watch over the stream and making immediately available sand bags and the resources to deploy them. More outbreaks occurred in the stream valley and at the side of the stream, which the drill company dealt with according to the contingency plan (Figure 20.8). When the outbreaks appeared in the middle of the stream and could not be dealt with, the EA demanded a new solution.

The drill company drilled a borehole at the side of the stream down to the top of the HDD bore. This let the grout leak preferentially up the casing into a prepared lagoon (Figure 20.9). From this lagoon, a pump transferred the grout up the valley, across the road and into the HDD reception pit.

Figure 20.8 *Second breakout controlled with sand bags in stream*

Eventually the quantity of grout emerging up the casing reduced so that the flow consisted mainly of groundwater. The lagoon was then used to settle the cloudy water and clean top water discharged through straw bales directly into the stream (Figure 20.10). The contractors monitored the stream and completed a daily observation log. There were no more outbreaks of grout.

It was agreed that the problem arose from a lack of information on the nature of the underlying rock, which appeared to be more fissured than had been thought. The Agency reported that it had experienced similar problems in the past from another HDD in the area. Client, contractor and drilling contractor all avoided prosecution by co-operating closely with the Environment Agency and by deploying significant resources and funds to solving the problem.

Figure 20.9 *Grout emerging from borehole into lagoon*

Contractor's learning points

1. Thorough planning and site investigation are essential.
2. Experiences of similar operations in the area is useful.
3. Early consultation with the Environment Agency is recommended.
4. Rigorous monitoring and emergency procedures are necessary to prevent pollution.

Figure 20.10 *Clean discharge from lagoon entering stream through straw bales*

20.5 OPEN EXCAVATIONS AND DIVERSIONS

When excavations in water are required, the area is generally cut off from the water by one of the following options:

- clay bund
- sand bags
- stop planks
- cofferdams (using sheet piles, diaphragm walls etc)
- caissons
- specialist dams (fabric, inflatable etc).

Major works are, by their nature, thoroughly planned and closely controlled. The Construction (Health, Safety and Welfare) Regulations 1996 require that "cofferdams and caissons are properly designed, constructed and maintained". Pollution prevention should be an integral part of planning and undertaking any works within water.

Clay dams and sand bags are most suitable for small and/or temporary works. Clay bunds can cause water pollution on removal. Placing straw bales downstream of the works can help filter out any silts that are generated. For small watercourses, sand bags, clay bunds and flume pipes (Figure 20.11) are used in a similar way to establishing the haul route crossing in Figure 20.4. If the flow is high, the water should be overpumped from upstream to downstream to prevent it backing up (see Section 20.5 for more information on overpumping). Whereas sand bags and bunds may be applied to smaller diversions, larger-scale works require dedicated silt-retaining measures such as those illustrated in Figure 20.12.

Figure 20.11 *Flume pipes carrying water flow during open-cut crossing of stream for pipeline installation (courtesy Transco)*

Stop planks and sheet piles must be installed correctly to avoid water ingress into the works area reaching unacceptable levels (ie so that the water needs to be pumped out). Use of "silent" piling should be considered to avoid fish-kill when piling over or adjacent to water. The tops of temporary sheet piles should be left above the maximum water level of the watercourse to cater for flood flows. Damage can occur to the impermeable lining of canals and certain other watercourses when stop planks or piles are removed, so these techniques should be avoided in such circumstances unless the piles or stop planks can be cut off at bed level.

Figure 20.12 *Silt trap comprising bags of sand used during watercourse diversion (courtesy Balfour Beatty)*

Cofferdams and caissons may be constructed from concrete or used to contain concrete operations. Normally the forms for the construction works will be provided by precast sections or sheet piles. Whichever method is used any joints need to be properly sealed and clutches on sheetpiles properly engaged to prevent fines polluting watercourses or groundwater. See Chapter 22 for guidance on using concrete in or near water.

Specialist dams for use in large watercourses that consist of a fabric curtain hung from a string of buoys are commercially available and easily portable. They can be supplied with flume pipes to maintain water flow. They can be subject to both undercutting and, as they do not extend far above water level, boat impact. Inflatable dams are also commercially available. When specialist equipment is being used the supplier's guidelines should always be followed, in consultation with the environmental regulator.

Bed profiling before and after works is often required to confirm that construction debris has been removed and there are no changes to the bed that will affect the hydraulics of the watercourse. For further guidance on bank reinstatement see Section 20.7.

New channels and diversions should be designed and constructed with consideration for pollution and ecology as well as engineering suitability (see Figure 20.13 and the case study below).

Figure 20.13 *River diversion to allow open-cut crossing. Lined with geotextile to avoid silt generation (courtesy Entrepose)*

> **Key guidance**
>
> Environment Agency (1999). R&D Publication No 11 *Waterway bank protection: a guide to erosion assessment and management*
>
> Fisher and Ramsbottom (2001). *River diversions: a design guide*
>
> Morris and Simm (2000). *Construction risk in river and estuarine engineering*
>
> River Restoration Centre (2002). *Manual of river restoration techniques*

> **Case study – diverting and culverting watercourses on road scheme** (source Balfour Beatty)
>
> Over the length of its route, the A120 crosses many watercourses, seven of which have Grade A status. Various improvements were considered during design and construction of the bridges and culverts:
>
> - straightening culverts to reduce the amount of concrete used and increase the levels of light within the culvert
> - use of soft engineering materials for all the watercourse diversions – eg putting down seeded matting where possible rather than using hard engineering methods such as gabion baskets, which can prove costly and do not promote vegetation regrowth
> - incorporating mammal shelves above the flood level on all culverts to allow mammals such as otters to pass under the new road during flood conditions,
> - including, in certain culverts, low-flow channels to increase the depth of water flowing through the culverts all year round
> - incorporating erosion protection measures at the outlet and inlet to the culvert to ensure smooth hydraulic transition and to avoid erosion
> - building oversized culverts where possible to allow for severe flood conditions
> - constructing pools and riffles within the new diversions with the help of the EA to create natural bed features.
>
>

20.6 OVERPUMPING

Short-term works (ie lasting no more than one day) across the whole width of a watercourse, particularly in still or slow-flowing water, can be undertaken by simply damming off the works area. When works are likely to take longer, and/or the flow is high, overpumping will be required to maintain the "flow" of water from upstream to downstream of the works without flooding the site.

The flow and watercourse depth should be monitored during wet and dry conditions before work starts. The pumps need to be of an adequate size to carry flow in wet or high-tide conditions. The outfall pipe should be placed well downstream of the works and protection, such as large stones or geotexile matting, provided to avoid scouring of the bed and/or banks at the outfall (see Figure 20.14). Further guidance on outfall protection is provided in Section 19.8.2.

If water has collected in the working area at the start of the shift, personnel should avoid entering or disturbing the area before it can be pumped out, to minimise silt.

Figure 20.14 *Large stones used for bed protection at a pipe outfall*

20.7 BANK WORKS

Although some works require "hard" bank protection, "soft" techniques, which allow vegetation to establish at the water's edge, should be employed wherever possible. Leaving the ground surface broken up to revegetate naturally is a simple, cost-effective method, as used in the reinstated river crossing illustrated in Figure 20.15. Rapid stabilisation for areas prone to erosion can be achieved by placing biodegradable matting (hessian, coir etc) and seeding it with fast-growing grasses. Other "soft" techniques available include reed-planted coir fibre rolls and brushwood rolls. Timber washboards (Figure 20.16), gabions and stone-filled mattresses, pitch and dry stone walls are also appropriate.

If there is a need to plant new shrubs or trees, the use of fertiliser should be avoided where possible or, is essential, carefully controlled during planting. Any herbicide to be applied prior to planting or during maintenance should be suitable for use near water and only used with the environmental regulator's permission.

Where protected species have been identified, such as water voles or white-clawed crayfish (Chapter 24), their habitat must be preserved within any bank protection. Any soil, plants and aquatic vegetation that have been removed will need to be reinstated.

The walls of canals and some channelled watercourses are not structural retaining walls, but act merely as erosion protection. They are often hundreds of years old. Construction plant should be lightweight, or restricted from approaching the edge of a canal.

Where construction activities involve work on existing flood defences, temporary arrangements need to be made to ensure that flood defence capacity is not compromised.

Figure 20.15
Reinstated open-cut river crossing, with no hard protection, to allow revegetation (courtesy Entrepose)

Figure 20.16
Timber boards and coir matting installed as river bank stabilisation (courtesy Transco)

> **Key guidance**
>
> Environment Agency (1999). R&D Publication no 11, *Waterway bank protection: a guide to erosion assessment and management*
>
> River Restoration Centre (2002). *Manual of river restoration techniques*
>
> Application form WQM1 for herbicide use from Environment Agency National Customer Contact Centre; tel: 08708 506506.

20.8 WORKS NEAR WATERCOURSES

Works alongside or near to watercourses may need consent from the environmental regulator to protect flood defence and water pollution interests.

Surface runoff is the most significant risk because of the short distance from the works to the watercourse. Chapter 18 gives detailed guidance on runoff control, but the following points are specifically relevant to work adjacent to watercourses. Before existing vegetation is removed, a buffer strip should be left along the edge of the watercourse and/or around the works to help filter any silty runoff (see case study below).

> **Case study – topsoil strip for pipeline adjacent to watercourses** (source: Alfred McAlpine)
>
> Constructing large-diameter cross-country pipelines involves large areas of stripped ground near numerous watercourses. One technique for minimising the impact of large volumes of silty runoff caused by rainfall is to stop the topsoil strip short of the watercourse crossing. This leaves a strip of vegetated topsoil across the spread protecting the watercourse. The resultant step (up to 40 cm high) holds water back and provides a face for the water to soak into and vegetation for the silty water to run over should the water reach that level.
>
> In light to medium rainfall events this technique proved very successful. During prolonged heavy rainfall it was overwhelmed.

Earth bunds or cut-off ditches should be constructed around site compounds and other works to isolate them from the water body. If necessary, runoff can be channelled to a settlement area before it is released to the watercourse (with permission). Where volumes are small, runoff should be directed over grassland where possible. If the ground slopes down towards the watercourse, runoff volumes will be greater. In these cases, earth bunds should be used along the length, or a cut-off trench dug at the lowest point to direct runoff away to a suitable area for treatment by settlement etc (see case study below). Filling the trench with gravel will help filter out silt and oil.

Case study – runoff from steep haul route down to watercourse

A temporary access track down a steep valley to a watercourse was constructed to manage an unplanned incident at the watercourse. The track was stripped down to the clay subsoil and the topsoil stored at the side. This generated silty water every time it rained.

The solution was to construct low earth bunds across the track at intervals to divert the runoff off the track, through the topsoil mound and into the adjacent field. Runoff was filtered through straw bales and geotextile to remove sand particles.

Case study – construction alongside a sensitive watercourse

Construction involved the installation of a pipeline alongside a stone-walled watercourse that fed a potable water supply reservoir. The topsoil was peat with clay underneath. The landscape was characterised by steep slopes and the rainfall was high.

The solution was to minimise the working width drastically, use excavated soil as a barrier to prevent runoff into the watercourse, minimise the time that the subsoil was exposed by laying short lengths of pipe, fluming the watercourse in some locations to isolate it from the works and deploy a permanent oil-absorbent boom downstream of the operation.

These measures were successful in preventing both silt and oil pollution of the watercourse and reservoir.

When working adjacent to water care should be exercised when slewing concrete skips or mobile concrete pump booms over open water.

20.9 WORKS IN THE FLOODPLAIN

Works in the floodplain should be completed in the short possible timeframe. As far as reasonably possible, they should all, including temporary works, be designed for flood conditions. Those preparing detailed flood emergency and contingency plans should:

- identify flood-prone areas from the environmental regulators or refer to guidance such as Planning Policy Guidance Note 25 *Development and flood risk* (DTLR, 2001), <www.odpm.gov.uk>
- schedule the construction to avoid works in flood-prone areas during the winter
- sign up for fax or email forecasts from the Met Office's MetBuild Direct service at <www.met-office.gov.uk/construction/mbdirect/index.html>
- register for emergency flood services and automatic telephone updates (where available):
    - ❖ <www.environment-agency.gov.uk/floodline>
    - ❖ <www.sepa.org.uk/flooding/index.html>
    - ❖ <www.riversagencyni.gov.uk/rivers/floodemergency-whotocontact.htm>
- check local radio and television forecasts and updates.

See Chapter 15 and Sections 18.4 and 18.4 for further guidance.

Permission must be obtained from the environmental regulator to ensure that the design and operation of the works in the floodplain is not likely to increase the potential for flooding or create a risk of flood damage. Where construction activities involve work on existing flood defences, temporary arrangements need to be made to ensure that flood defence capacity is not compromised.

20.10 WORKS OVER WATER

Cleaning and painting processes over or adjacent to water have the potential to pollute. Where possible, physical cleaning methods (eg wire-brushing, sand- or grit-blasting) should be adopted in preference to the use of liquid chemicals, such as caustic and acid solutions. If such liquids have to be used then the effluent must be fully contained by the use of a bund or tray. It is important to establish that the paint and chemicals to be used are harmless to aquatic life.

Surface preparation should be carried out in a controlled manner, under sheeted enclosures to prevent dust and coarser materials from falling into the water and to make it possible to collect and dispose of those materials (see Figure 20.17). Shot-blasting operations should be carried out behind screens to prevent dust escaping. If dust is dampened down with water sprays, the runoff should be contained and not allowed to drip into the water.

Figure 20.17 *Sheeting of Pontcysyllte Aqueduct over River Dee SSSI (courtesy British Waterways)*

When constructing bridges or similar structures, care should be exercised when slewing concrete skips or mobile concrete pump booms over open water (Figure 20.18). The guidance in PPG5 *Works in, near or liable to affect water* (EA, SEPA and EHS, 2000) should be followed.

Figure 20.18 *Beams for the M1A1 bridge being erected over the River Aire (courtesy Balfour Beatty)*

Checklist – works in or near water

1. At an early stage discuss the works and obtain consent from the environmental regulator and, where necessary, British Waterways and/or the internal drainage board.

2. Assess the pollution risks of the works and mitigate the risks at source wherever possible.

3. Prepare an emergency plan and include pollution control measures in all method statements. Make sure staff are aware of the contents of these documents.

4. Monitor water quality before, during and after works.

5. Monitor the water flow and depth during wet and dry conditions before work starts. Size any pumps and culverts adequately to carry flow in flood or high tide conditions.

6. Avoid entering watercourses: provide a temporary bridge or culverted haul road access.

7. When stripping topsoil, leave a grass buffer strip next to watercourses to filter runoff.

8. Install sediment filters or traps downstream of the works.

9. Take particular care when using concrete adjacent to, over or within water.

10. Site diesel pumps and static plant in drip trays as far away from the watercourse as possible. Refuel plant way from the water. Keep a stock of oil-absorbent materials nearby. Fix an oil-absorbent boom across the watercourse downstream of the works.

11. Where overpumping or dewatering, place the outfall pipe well downstream of the works and protect the bed and/or banks from scouring.

12. At the end of the shift ensure any pumps have sufficient fuel to run overnight if necessary. If water has collected in the working area at the start of the shift, avoid entering or disturbing the area before it can be pumped out, to minimise silt

21 Excavations and dewatering

Dewatering is the process by which water is removed either from the ground or from within an excavation. Dewatering covers a range of situations and processes, but is typically applied when:

- the water table is lowered to allow a wide or deep excavation to take place in "dry" conditions; dewatering can take place within or outside the excavation
- an excavation extends below the water table and groundwater seeps in, requiring removal
- rainfall or surface water runoff has to be removed from open excavations
- works need to take place in water and the area is sealed off and pumped out, eg a cofferdam.

Successful dewatering is dependent not only on getting rid of existing water but also on stopping any more getting in. Preparation and clear planning pay dividends.

21.1 LEGAL REQUIREMENTS

A transfer licence will be required under the Water Act 2003 (in England and Wales). Authorisation (in the form of registration, general binding rules or a licence) may be required under the Water Environment (Controlled Activities) (Scotland) Regulations 2005 for larger dewatering operations. Regulatory approvals will include the dewatering as well as the disposal of the dewatered water. Operations should be discussed with the environmental regulator at an early stage so that the requirements for any authorisation can be agreed, and so the licence itself can be processed (which can take up to **four months**).

21.2 EXTERNAL DEWATERING (GROUNDWATER)

Well-point and deepwell dewatering systems can be used to overcome saturated and unstable ground. Well-points are a series of interconnected small filter pumps placed in the subsoil to a predetermined depth that can achieve shallow drawdown of the water table (generally 4–6 m). Deepwell systems are installed in boreholes to achieve drawdown at greater depths.

Before any dewatering to lower the water table takes place, the environmental regulator must be consulted so it can issue an appropriate authorisation. This is likely to set limits on the volume and rate of dewatering (Chapter 13).

Any groundwater abstracted will need to be discharged. It is widely accepted that the optimum solution is to return it to groundwater (groundwater recharge). Alternatively, it will need to be discharged to a watercourse, surface water drain or, least preferable in the case of clean water, disposed of to a foul sewer. Advice should be sought before any recharge or discharge scheme is undertaken, because authorisation may be required (Chapter 13). The water quality must meet any conditions attached to a consent. Chapter 19 provides further guidance on water treatment and discharge options.

Generally, discharged water from a dewatering system will be clean, as it is drawn from the water table below ground. However, in the initial set-up and drawdown there is likely to be potential for discoloration, silt and other potential contamination, so it is essential to monitor the quality of the discharge closely. Where this is likely, water should be routed via a settlement tank or pond before it is discharged under the relevant consent from the environmental regulator.

If the ground on site is contaminated, the dewatered water is also likely to be contaminated and must be treated to an appropriate standard. There is also a higher risk of mobilising ground contamination from off site when working with groundwater. Before starting work, the level of risk for contaminated ground or groundwater should be investigated (follow the guidance in Chapter 23).

Dewatering groundwater may affect adjacent surface watercourse levels and ecology (Chapter 24). When dewatering over linear features, the impacts on the groundwater (and surface water) regimes are likely to be more varied and widespread. Water levels in surface watercourses and boreholes around the site should be monitored, particularly in sensitive areas. Chapter 14 provides guidance on monitoring. If water levels drop to unacceptable levels, one solution is to recharge surface waters or groundwater with extracted groundwater. The water needs to be of the correct quality and temperature – holding it in a settlement lagoon can assist in this. Another important consideration with dewatering is the risk of ground settlement and this too may need to be monitored.

Further illustrated guidance on groundwater control techniques can be found in CIRIA C532 *Control of water pollution from construction sites* (Masters-Williams *et al*, 2001) and C515 *Groundwater control – design and practice* (Preene *et al*, 2000).

> **Checklist for action – groundwater dewatering**
> 1. Do not dewater without previously consulting the environmental regulator.
> 2. Adhere to any conditions set by the environmental regulator on the volume and rate of dewatering.
> 3. Identify the disposal route for the dewatered water and seek agreement from the environmental regulator.
> 4. Monitor and treat dewatered water to a suitable quality before discharging it.
> 5. Monitor any ecological impacts such as lowering surface water levels.

21.3 INTERNAL DEWATERING (EXCAVATIONS)

21.3.1 Minimising water ingress

The first priority in controlling dewatering is to minimise the water ingress into excavations. There are several engineered groundwater exclusion techniques including:

- sheet piles
- diaphragm walls
- contiguous bored piles
- vibrated beam walls
- cut-off walls
- grouting.

Setting up these exclusion systems can itself affect water if there is a need to drill through aquifers, groundwater and contaminated ground, or to use grouts both above and below ground. Follow the guidance on groundwater (Chapter 2) and use of concrete and grout (Chapter 22). See C515 *Groundwater control – design and practice* (Preene *et al*, 2000) for further guidance.

Surface water runoff should not flow into excavations. Water running down the side of an exposed face may dislodge fine particles and take them into suspension. If there is water in an excavation, neither plant nor members of the workforce should be allowed to move about in it and stir up mud and silt. Fine particles such as silt and clay take a long time to settle out, so this can be an expensive process. Water should be diverted by digging cut off ditches around the excavation, grading the ground or placing sand bags or a small earth bund around the edge of the excavation.

21.3.2 Pumping out water

It is important to plan any pumping system well. Any pumped water will need to be disposed of to a grassed area for infiltration, a watercourse, a surface water drain or to a foul sewer without causing pollution (see Figure 21.1 and following case studies).

> **Case study – silt pollution offences** (source: *The ENDS Report*, Aug 2003)
>
> A construction company was fined £2000 and had to pay costs of £1223 after admitting to polluting a tributary of the River Cam. The court heard that silt entered a brook after the company pumped out a hole filled with silty water on to a public highway. The water entered the watercourse through roadside drains.

Figure 21.1
Discharging water from pumping activities has caused silt pollution in the stream

Advice should be sought at an early stage, as permission to discharge water may be required (Chapter 13). A "permit to pump" procedure can be set up whereby staff are not allowed to use a pump on site until a permit to pump form (see Figure 21.2) has been completed. The form should state the following:

- why the pumping operation is necessary
- where the water is being pumped to
- who is responsible for pollution mitigation measures
- what treatment is necessary before discharge
- whether a consent to discharge has been granted.

This procedure is not a legal requirement, but contractors often use it as part of their own site management. Before discharging any water, it is important to check that the appropriate permissions are in place and the water quality complies with any conditions. Where the water is silty or otherwise polluted, it should be routed via a settlement tank or pond before it is discharged. Chapter 19 provides more guidance on water treatment and disposal options.

> **Case study – fine for polluting wildlife site** (source: *The ENDS Report*, May 2004)
>
> The director of a construction company has been fined £3000, and his company £2000, after polluting a river designated as a special area of conservation under the Habitats Directive. They were also ordered to pay costs totalling £525.
>
> The contractor was renewing a culvert from which water discharges into the River Kent. The river is designated an SAC because it supports a population of white-clawed crayfish.
>
> Environment Agency officers on duty in the area saw that water was seeping into the hole dug to repair the culvert. A pump had been installed to stop the hole flooding. The officers told the builders not to turn on the pump as it would empty silty water into the river via the culvert. These instructions were ignored, however.
>
> The contractor thought it was acceptable to use the pump because the silt would settle out of the water in the culvert due to its shallow gradient. However, samples from the river showed that the incident discharged high levels of suspended solids into the river.

Figure 21.2 *Permit to pump (courtesy Alfred McAlpine)*

Using submersible pumps can generate more sediment through water turbulence. a corner of the excavation should be used as a sump and care taken to avoid disturbing that corner. Simple additional measures can be taken to reduce unnecessary sediment generation such as placing the pump in a perforated oil drum, a short length of wide-bore perforated pipe or concrete manhole rings containing granular fill. Alternative methods of holding the pump off the base of an excavation or body of water are illustrated in Figures 21.3 and 21.4.

Water in an excavation that remains open for some time can be controlled and channelled by granular-filled edge drains around the inside perimeter of the excavation, leading to the sump(s).

> Make sure an adequately sized pump is installed – remember that water volumes can increase significantly in wet weather.

Figure 21.3
Use of a ladder to keep the pump off the base of the excavation to avoid disturbing sediments

Figure 21.4
Use of an excavator arm for the same purpose

It is important to seek advice from the pump supplier when selecting pumps. Each pump has a performance curve and the resultant flow is dependent on the specific conditions encountered on site. The pumping capacity of one 6 in (152 mm) pump is equal to four 3 in (76 mm) pumps, not two, as might be expected. Most diesel-powered pumps have higher outputs than electro-submersible pumps of the same size. It is wise to check output and head when choosing pumps and not to rely solely on branch size.

The following should be considered when selecting a pump:

- length and lift of the suction pipe – the latter should be kept as low as possible
- length and height of the discharge pipe – discharge pipes should be generously sized (a rule of thumb is one size larger than the pump discharge branch) particularly on long-distance runs
- percentage of silt to water
- hydraulic or air-powered pumps
- presence of explosive atmospheres (contaminated water or foul drainage)
- type of hose or pipe to be used (eg flexible PVC, rigid plastic or steel).

For further guidance on the selection of pumps see CIRIA Report 121 (Prosser, 1992).

Uncontrolled water leaks from pumps and hoses often create additional surface water problems. Careful thought should be given to site access when setting up both the pump and hose run. Site plant can cause irreparable damage if it runs over a rigid hose. Lay-flat hoses may be more resistant, yet could still be damaged. To avoid damage, discharge hoses need to be routed out of the way of vehicle movements. Wherever hoses pass over a solid edge (the top of an excavation or a concrete sump, for example) care should be taken to ensure no damage can occur. Regular – even daily – checks should be carried out on the pump, hoses and couplings for leaks and kinks, with any problems being fixed immediately.

Electric pumps should be used wherever possible to reduce the use of fuels on site. These will require a starter control panel and float controls. Larger pumps require soft-start systems to ensure smooth starting from the mains or power generator supply. The pump supplier will be able to advise on the correct size of generator needed.

Diesel pumps should stand in a drip tray, and on hardstanding where possible. Noise pollution must be avoided, particularly if pumps are running 24 hours a day or are turned on early in the morning. Care in siting the pump and erection of a simple screen, earth bund or even parked-up plant can help reduce noise levels considerably.

21.3.3   Recharging excavations

Once works in water are completed, or when dry conditions in excavations below groundwater are no longer required, water should be allowed back into excavations or dewatered areas in as slow and controlled a way as possible. While removing upstream features, downstream bunds or cut-off structures should be left in place where possible. Guidance on works in water can be found in Chapter 20.

> **Checklist for action – excavation dewatering**
>
> 1. Minimise water ingress into excavations by engineered means or simple cut-off drains or sand bags around excavations.
> 2. Identify a disposal route for dewatered water and seek agreement from the environmental regulator or sewerage undertaker early on.
> 3. Use a permit to pump system and select a pump to cope with wet weather conditions.
> 4. Set up drains and sump pumping surrounded by granular fill to avoid generating sediment.
> 5. Check pumps and hoses for leaks and fix as soon as possible.
> 6. Site pumps on drip trays.

22 Concrete and grouting activities

Concrete, bentonite, grout and other cement-based products are highly alkaline and corrosive and can have a devastating effect upon water quality. Cement-based products generate very fine, highly alkaline silt (11.5 pH) that can physically damage fish by burning their skin and blocking their gills. This alkaline silt can also smother vegetation and the bed of watercourses and can mobilise pollutants such as heavy metals by changing the water's pH. Concrete and grout pollution is often highly visible.

Large volumes of cementitious materials are used within the construction industry and frequently involve the batching and placement of "wet" materials in situ or the use of cementitious "muds" for drilling and grouting. Particular risks are posed to water quality when construction is taking place over or near surface waters (eg bridges or headwalls) and underground (eg tunnelling, directional drilling or below-ground structures).

22.1 LEGAL REQUIREMENTS

It is an offence under the Water Resources Act 1991 (in England and Wales), the Water Environment and Water Services (Scotland) Act 2003 and the Water (Northern Ireland) Order 1999 to discharge any polluting material, including cement-based products or water polluted with concrete, grout etc, into any surface water body, groundwater or surface water drain without prior consent.

The use and storage of cement-based products must be carried out in accordance with COSHH requirements and the manufacturer's data sheets (see also Chapter 16).

> **Case study – concrete pollution costs contractor £4500** (source: *The ENDS Report*, Apr 2004)
>
> A company that claims to be "one of the foremost contractors" in the UK's pipeline and utility construction sectors has paid £4500 after killing about 100 trout.
>
> Investigating officers found dead and dying brown trout in the River Calder, Lancashire. The fish had lost their mucus covering and had burns to the eyes and fins. The company was carrying out construction work on a pipeline crossing beneath the river. The contractor had obtained a land drainage consent from the Agency and been advised on the need to prevent pollution. It agreed that water coming into contact with the wet concrete casing of the pipe should not be discharged to the river, but allowed to settle and then filter through the ground. However, the pump had been turned off for several days and water that had become extremely alkaline through contact with the cement was discharged to the river. The contractor was fined £2000 and ordered to pay costs of £760. It had already paid £1753 to the Agency to reimburse its investigation costs.

22.2 ALTERNATIVE METHODS

Where the site is within or near a particularly sensitive water environment, alternative construction methods should be investigated. For example, use of pre-cast or permanent formwork will reduce the amount of in-situ concreting required above, within or adjacent to water. In sensitive environments, ready-mix suppliers should be used in preference to on-site batching.

For work below the water table or within water, hydrophilic (water-repelling) grout and quick-setting mixes or rapid hardener additives ought to be used where appropriate, to promote the early set of the concrete surface exposed to water. When working in or near water, and application *in situ* cannot be avoided, the use of alternative materials such as biodegradable shutter oils should be considered.

22.3 ON-SITE BATCHING

Production activities may vary in scale from fully automated concrete batching plants on large projects to hand-mixing of mortar for small works. Regardless of the size of the operation, cementitous material must not be allowed to enter the water environment.

> **Checklist – on-site batching**
>
> 1 Locate batching and mixing activities and material storage areas well away from watercourses and drains.
> 2 Control surface drainage in the area around the batching plant, as it may be polluted.
> 3 Do not hose down spills of concrete, cement, grout or similar materials into surface water drains.
> 4 Washout from mixing plant or concrete lorries should be carried out in a designated, contained impermeable area (see Section 26.4.1).
> 5 Clean and recirculate water in batching plants as much as possible, where water is

22.4 TRANSPORT AND PLACEMENT

Concrete, bentonite and grout may be transported by truck mixers, tipper wagons, site dumpers, machine buckets and skips. Various methods can be used for placing concrete, bentonite and grout, including pumps, skips and by injection. It is vital to plan and supervise all activities and to adopt special procedures, in particular where there may be a risk of pollution to surface water or groundwater.

> **Planning is essential to minimising the potential for pollution**
>
> 1 When working below ground, identify potential pathways for concrete or grout loss such as geological fissures.
> 2 Discuss arrangements for deliveries to site with suppliers before work starts, agreeing routes, designated washout areas and emergency procedures, particularly where there are several remote locations.
> 3 Develop procedures and a contingency plan for monitoring the use of cement-based products and uncontrolled releases.
> 4 Train all site personnel on method of work and emergency procedures.
> 5 Establish clearly signed designated washout areas (see below).

22.4.1 Washout areas

A frequently overlooked area of water management is the washing out of concrete wagons (or other plant used in concreting and grouting operations). One or more designated contained washout areas should be provided, located away from watercourses, drains and protected groundwater zones. They should be impermeable, to prevent pollution of groundwater. Washout areas should be clearly signposted and delivery drivers should be informed of their location (and operation if necessary).

Washout water should be contained in a pit, skip or series of settlement pits, depending on the volumes expected. For a one-off concrete pour, a skip or lined pit may be suitable. For longer-term concrete operations, or for an on-site batching plant, a series of lined ponds would be suitable. At a batching plant, the ponds can also be used to contain runoff from the forecourt of the batching plant. Excess concrete may be placed on to a suitable area of ground (it can subsequently be broken up and reused) to avoid unnecessarily filling the washout area with concrete.

Where work is taking place over an aquifer or within a source protection zone (see Chapter 2), concrete washout water must be contained and removed from site for treatment as liquid waste.

Where possible use should be made of water from settlement ponds or similar facilities, or recycled washout water. Where concrete is delivered to site, only the chute need be cleaned, using the smallest volume of water necessary, before leaving the site. Wagons should wash out fully back at the batching plant, where facilities will be provided. Operatives should be instructed to employ a triggered hose and to use the minimum volume of water necessary. If it is not practical to wash just the chutes (for example, if the delivery area is far from the batching plant), the washout area ought to be designed to take a larger volume of water.

Washout areas require maintenance. Once the solids have settled out, and if chemically suitable, the water can be recovered for use in subsequent batches of concrete or may be reused to wash out trucks to minimise the volume requiring disposal. Settled silt, surplus wet concrete and hardened concrete will have to be removed. The removal frequency will depend on the use of the washout area, and can range from weekly to just on completion of concreting. If it is a suitable consistency, it may be possible to mix this material with excavated spoil going off site. Hardened concrete will need to be broken out and could potentially be reused as aggregate on site. Liquid silts will require pumping out by a specialist contractor and must be disposed of in accordance with waste legislation.

Because of its high pH (alkalinity), washout water may not be suitable for discharge to surface water drains and may even need treatment before disposal to the foul sewer. Further advice on treating concrete-polluted water is provided in Section 19.3. In all cases a licence or consent will be required (see Chapter 13).

Case study – concrete washout ponds in series (source: Hyder)

Concrete operations on a major tunnelling project involved on-site batching and continuous delivery for several months. A dedicated washout area was constructed. Washout water and small amounts of surplus concrete were discharged into the top pond, where solid material settled out. When full, the water overtopped into a second and then a third pond. When the third pond filled up, a known volume of water was pumped into an *in-situ* treatment tank and dosed with a powdered acid to reduce the pH level. The pH was monitored and recorded before the water was released into the sewer under a consent from the local sewerage undertaker.

Series of settlement ponds with treatment tank in bottom left corner

22.4.2 Underwater concreting

There should always be close liaison with the environmental regulator to agree a suitable method for underwater concreting.

Where concrete is to be placed under water or in tidal conditions it will need to be designed to provide a cohesive mix to limit segregation and washout of fine material. This will normally be achieved by having either a higher than normal fines content, a higher cement content or the use of chemical admixtures.

Most underwater concrete will be placed within the confines of a cofferdam or caisson. Normally, the forms for the construction works will be provided by pre-cast sections or sheetpiles. In either case, it is essential to seal joints securely and to engage clutches on sheetpiles properly to prevent fines polluting watercourses or groundwater.

Plant will also be operating close to water, but special consideration should be given to the transport of concrete from the point of discharge from the truck-mixer to final discharge into the delivery pipe (tremie). Care should be exercised when slewing concrete skips or mobile concrete pump booms over open water and contingency plans should be in place in the event of the concrete pump pipework becoming blocked.

Attention needs to be paid to the eventual dewatering of the construction, as this may be more alkaline than the groundwater and will contain some washout from the placing of the concrete. See Chapter 21 for more information on dewatering.

22.5 TUNNELLING, THRUST-BORING AND PIPEJACKING

Tunnelling, thrust-boring and pipejacking (directional drilling) are used as trenchless techniques for installing underground pipelines, ducts, culverts and services. They are specialist activities and the size, nature and risk of the work need to be assessed site by site. If a pathway exists, these activities may pollute groundwater and surface water. See Section 20.4 for more information on directional drilling under or near watercourses.

It is essential to conduct a comprehensive site investigation to ensure that ground conditions are acceptable for the selected method of work, and also to minimise the risk of water pollution. A risk-based approach should be adopted, with particular consideration given to the geology of the area and the water table.

Potential polluting materials that are often used during the process include:

- bentonite – a bentonite slurry is used to provide a slip coating when pipejacking to reduce friction on the ground. It is injected into the annulus around the pipe either through the shield or through holes in the wall of the pipe
- grout – used to complete the installation of pipes or tunnel segments and to infill void space and overbreak
- other conditioning agents and additives – for example tunnelling foams, may also be injected into the ground to aid lubrication and/or maintain the earth pressure balance during tunnelling. The risk of pollution can be reduced by seeking biodegradable alternatives from suppliers.

It is important to assess the quantity of material required before carrying out the task. Having done so, experienced operatives can monitor the quantities being used against the estimated volumes and so are swiftly alerted to any grout loss. All watercourses nearby should be monitored while the works are in progress. Material stockpile and storage areas need to be designed to allow for the collection and management of runoff wet tunnel or drilling arisings (see key guidance below).

> **Key guidance**
>
> CIRIA, on behalf of Pipejacking Association (in prep). *Slurry: management and disposal of semi-liquid spoil*
>
> Pipe Jacking Association (1995). *Guide to best practice for the installation of pipe jacks and microtunnels*
>
> Pipe Jacking Association (in prep). *Management of process arisings from tunnels and other earthworks: a guide to regulatory compliance*
>
> Pipe Jacking Association, <www.pipejacking.org>

> **Concrete, bentonite and grout**
>
> 1. Check formwork and shuttering before pouring to ensure it is stable and joints are sealed to prevent loss.
> 2. Monitor and record the use of cement-based products during each operation.
> 3. Where possible, prevent concrete skips, concrete pumps and machine buckets from slewing over water while placing concrete.
> 4. Secure the end of pump hoses by means of a rope during concreting over and adjacent to waters to prevent discharge hose accidentally depositing concrete away from the pour site.
> 5. Take care when slewing concrete skips or mobile pump booms over open water.
> 6. When placing concrete by skip, securely fasten the opening gates of the delivery chute by lock chain to prevent accidental opening over water.
> 7. Construct earth or sand bag barriers around bentonite and grout-mixing areas, supply lines and around working areas to prevent escape.
> 8. Position supply lines as far away from water and drains as possible.
> 9. Cover freshly placed concrete to avoid the surface washing away in heavy rain.
> 10. Clean up any spillages of cementitious materials immediately and dispose of correctly.

23 Contaminated land

23.1 INTRODUCTION

Part IIA of the Environmental Protection Act 1990 and Waste and Contaminated Land (Northern Ireland) Order 1997 defines "contaminated land" as land "that is actually, or could be causing, an unacceptable risk to human health and the environment, including controlled waters". Although all brownfield sites have the potential to be contaminated to some degree by historical activities, and even greenfield sites may be contaminated (by illegal tipping, for example), they do not necessarily constitute "contaminated land". Common examples of contaminating activities and associated hazardous substances are listed in CIRIA R132 *A guide for safe working on contaminated sites* (Steeds et al, 1996).

Land contamination provides a source of pollution that can be mobilised by on-site or adjacent construction activities (eg piling, dewatering, excavations), potentially leading to:

- costly and lengthy delays to large-scale construction projects
- severe pollution of surface water and/or groundwater
- serious impacts on water users
- significant damage to receiving natural environments (eg wetlands) and ecological systems
- prosecution.

23.2 INVESTIGATION AND ASSESSMENT

Early and detailed planning coupled with knowledge and risk-based assessment is essential to minimise the risk of construction activities on, or adjacent to, contaminated sites causing surface water and groundwater pollution.

The site investigation data and the ES (where available) should be consulted to enable known contaminated areas to be identified (see Section 8.4). Where specified in contract documentation, measures to deal with the contaminated land should be followed.

The findings of the investigative and assessment works for each contaminated site should be used to:

- design specific mitigation measures to manage contaminated soils, groundwater and runoff
- prepare detailed method statements to manage contaminated soils, groundwater and runoff
- determine appropriate *in situ* or *ex situ* remedial strategies (as necessary).

23.3 DEVELOPMENT OF SPECIFIC MITIGATION

Large-scale linear construction projects may encounter numerous individual contaminated sites, both within and adjacent to the route. Specific mitigation and management measures should be developed for each contaminated site along the route, as the contamination history, contamination characteristics and environmental setting (including geological, hydrological and hydrogeological sensitivity) of each location will be different. This will require significant specialist input and consultation with the environmental regulator.

Special emphasis should be placed on construction activities that present a high risk of generating pollution pathways on, or adjacent to, contaminated land. Common examples of high-risk construction activities are outlined in Table 23.1.

Table 23.1 *Construction activities with pollution risks from land contamination*

Construction activity	Pollution risks
Excavations and earthworks on contaminated land	Handling contaminated materials; generating contaminated runoff; creating vertical and horizontal pollution migration pathways
Stockpiling of contaminated soils and material	Generating contaminated runoff, discharge and dust
Tunnelling and excavations in contaminated land	Creating horizontal pollution migration pathways
Piling into contaminated land	Creating vertical pollution migration pathways
Dewatering on or adjacent to contaminated land	Drawing contaminated groundwater into on-site works; generating a contaminated discharge. Mobilising contaminants through changes in reduction-oxidation potential in the ground
Discharges to ground on contaminated land	Remobilising contamination within the unsaturated zone causing groundwater pollution
Aquifer recharge on or adjacent to contaminated land	Raising the water table and remobilising contamination within the unsaturated zone causing groundwater pollution

23.4 MANAGING UNEXPECTED CONTAMINATION

Land contamination can vary in scale from a couple of buried barrels of waste to entire industrial sites. Detailed planning and investigation should identify land contamination along the route, but there will always be a residual risk of encountering unexpected contamination.

Site operatives should be:

- made aware of the risk of encountering unexpected contamination
- trained to recognise potentially contaminated materials
- trained to implement contingency plans and procedures.

Special contingency plans and procedures, following best-practice guidance, should be developed for the unexpected discovery of contaminated materials. Example procedures are outlined below. For further guidance see CIRIA C650 *Environmental good practice (second edition)* (Chant-Hall *et al*, 2005a).

Example contingency procedures to manage the discovery of unexpected contamination

1. Stop work immediately.
2. Inform the site supervisor and site manager.
3. Ensure all site operatives are safe.
4. Isolate and contain the area and any excavated material (as is practicable and safe) – to reduce the risk of spreading contamination or generating contaminated runoff or dust.
5. Inform the environmental regulator.
6. Arrange for a technical specialist to conduct chemical testing and a risk assessment.
7. Revise and develop method statements and remedial strategies accordingly.

Key guidance

Assessment

Aspinwall & Co (1994). CLR1 *A framework for assessing the impact of contaminated land on groundwater and surface water, vols 1 and 2*

DoE (nd). "Industry Profiles", a series of guidance documents on possible contamination from various industries, available from Defra and EA

Rudland et al (2001). C552 *Contaminated land risk assessment. A guide to good practice*

Remediation

Barr et al (2002). C575 *Biological methods for assessment and remediation of contaminated land – case studies*

Barr et al (2003). C588 *Non-biological methods for assessment and remediation of contaminated land – case studies*

Evans et al (2001). C549 *Remedial processes for contaminated land – principles and practice*

Privett et al (1996). SP124 *Barriers, liners and cover systems for containment and control of land contamination*

Rudland and Jackson (2004). C622 *Selection of remedial treatments for contaminated land. A guide to good practice*

CIRIA (2005). SP164 *Remedial treatment for contaminated land Vols I–XII*

Groundwater

Holden et al (1998). R186 *Hydraulic measures for the control and treatment of groundwater pollution*

Marsland and Carey (1999). R&D Publication 20 *Method for the derivation of remedial targets for soil and groundwater to protect water resources*

Preene et al (2000). C515 *Groundwater control – design and practice*

Working on contaminated sites

Chant-Hall et al (2005a). C650 *Environmental good practice (second edition)*

Environment Agency (2001c). NC/99/73 *Piling and penetrative ground improvement methods on land affected by contamination – guidance on pollution prevention*

Steeds et al (1996). R132 *A guide for safe working on contaminated sites*

Checklist for action – land contamination

1. Do not start work on or near contaminated land until a competent professional has undertaken a comprehensive investigation, risk assessment and mitigation measures.
2. Divert uncontaminated surface water runoff away from contaminated areas.
3. Do not allow site drainage to mix with contaminated discharges or liquids.
4. Control and collect all contaminated liquids, eg contaminated groundwater from dewatering activities, runoff from contaminated areas.
5. Store all contaminated liquids (including runoff) in secure and bunded tanks on an impermeable surface, for appropriate on-site treatment and/or off-site disposal at a licensed facility.
6. Always minimise the volume of contaminated material generated, eg during excavations, dewatering or runoff collection.
7. Do not stockpile contaminated materials unless absolutely necessary.
8. If stockpiling is necessary, store the contaminated materials on impermeable or hard-standing areas, away from watercourses and uncontaminated materials.
9. Cover contaminated materials to reduce the generation of contaminated dust and runoff.

24 Ecology

New legal controls and contract conditions are increasing the level of protection being afforded to wildlife and natural features. Usually the developer is responsible for investigating the ecological interests and sensitivities of a site and defining them for the contractor. In many instances, though, it will be necessary to consult an environmental expert for specialist advice. If contractors fail to meet their legal and contractual requirements, sanctions may be imposed, which can affect the cost and programme for the project.

This chapter considers the effects on ecology of water pollution, rather than the effects of construction in general.

> **Key guidance**
>
> Chant-Hall *et al* (2005a). C650 *Environmental good practice (second edition)*
>
> Highways Agency (1998). *Design manual for roads and bridges*, Sections 10/a and 11
>
> Newton *et al* (2004a). C587 *Working with wildlife*
>
> Newton *et al* (2004b). C613 *Working with wildlife pocket book*
>
> Scottish Executive (2000). *River crossings and migratory fish: design guidance*, <www.scottishexecutive.gov.uk/consultations/transport/rcmf-00.asp>
>
> Venables *et al* (2000a). C512 *Environmental handbook for building and civil engineering projects. Part 1 Design and specification*
>
> Venables *et al* (2000b). C528 *Environmental handbook for building and civil engineering projects. Part 2 Construction*

24.1 LEGAL PROTECTION

Those who damage or disturb protected habitats or species can be prosecuted under a range of international, European and UK legislation. The fine for non-compliance with legislation varies according to the species and the type of damage caused. CIRIA C587 *Working with wildlife* (Newton *et al*, 2004a) provides an introduction to the principal aspects of law that protect wildlife in the UK.

Before working on any designated site in the UK prior written agreement must be obtained from the nature conservation body (see below); working methods will need to be agreed. It is an offence to start work without this permission. An environmental impact assessment may be required.

> **The term nature conservation bodies** is used here to represent the following organisations that have responsibility for promoting and regulating the conservation of wildlife and natural features:
> - Countryside Council for Wales
> - Natural England (from 2006; formerly English Nature, the Countryside Agency and Rural Development Service)
> - Northern Ireland Environment and Heritage Service
> - Scottish Natural Heritage.

24.1.1 Protected sites

Certain designated areas are afforded legal protection of various kinds. A summary of international, national and local designations is provided in Appendix 4. A linear project may encounter more than one site of designated ecological importance, particularly given that the impacts can be indirect (ie off-site or downstream) as well as direct.

Designated sites can be identified from a desk-based study, through consultation with the nature conservation body or from a review of the ES.

Written agreement from the relevant conservation body is usually required before work can begin in or near a protected site.

> **Case study – company fined for disturbing wildlife** (source: *The ENDS Report*, Dec 2004)
>
> A construction company was ordered to pay £7000 in fines and costs after disturbing freshwater wildlife habitats in Cumbria. The company admitted carrying out work on two sensitive tributaries of the River Mint without the necessary permission from the Environment Agency. The company was constructing pipelines that crossed several watercourses. Consultations with the Agency and English Nature alerted both stakeholders to the sensitivity of the area and the need for prior written permission from the Agency before any work was undertaken.
>
> The River Mint catchment is a candidate special area of conservation and site of special scientific interest because it supports bullheads, native crayfish and freshwater pearl mussels, as well as salmon and sea trout. Agency officers noticed that construction work affecting the becks was being carried out by the company without permission. The construction work disturbed the banks of both watercourses and vehicles were being driven through the water. No measures had been taken to stop silt polluting the streams. A crossing had been built for heavy vehicles, but the pipes used were positioned so high as to prevent fish passing through when the water levels were low.
>
> The court decided that the company had been negligent in failing to meet its environmental obligations when working in a highly sensitive location. The construction company pleaded guilty to three counts of wilfully disturbing waters where spawn or breeding fish may have been present. It was fined £1000 for each offence and ordered to pay costs of £4010.

24.1.2 Protected species

Certain species, and their breeding, feeding or resting areas, are afforded legal protection. Species most likely to be affected by water pollution include certain fish, newts and rare plants. The presence of such species makes the water environment even more sensitive to pollution. Other protected species, such as water voles, may be harmed or have their habitat destroyed by works to banks of watercourses.

Before construction work starts, the nature conservation body should be consulted, or the ES reviewed, to identify whether there may be protected species on the site. A qualified ecologist should be engaged to conduct surveys to identify protected species present both on the site and in the surrounding area (Figure 24.1). Some kinds of ecological survey may only be undertaken at certain times of the year and a licence may be required for certain species. Sufficient time needs to be allowed to obtain licences and carry out surveys. Appendix 3 provides guidance on survey timing. Further details on survey requirements, methods, timing and key species can be found in CIRIA C587 (Newton *et al*, 2004a).

In certain circumstances it is possible to obtain special permission to disturb protected species, either to catch them for translocation to another suitable habitat or to exclude them from the existing habitat. This must be done by a specialist individual or organisation under a licence issued by the conservation body.

Identify ecological constraints before starting construction work.

Plan works to avoid those times of year when works are legally restricted or prevented to avoid disturbance to protected species (see Appendix 3).

Figure 24.1 *Ecological survey of stream in SSSI before construction of a pipeline crossing (courtesy Hyder)*

24.2 CONSTRUCTION IMPACTS

The direct effects of pollution upon the surrounding ecology can be temporary or permanent. Pollution often takes the form of silty water: any construction work mobilising silt is potentially polluting and can cause long-lasting damage to river life. Silt can clog the gills of fish, suffocating them, and can smother spawning sites and insect habitats on the river bed, removing a source of food for fish. Silt can cover the leaves of aquatic plants, limiting their growth. Other effects on watercourse ecology include:

- changes in water quality, pH, salinity, turbidity
- oxygen depletion and/or nutrient enrichment, leading to algal blooms
- input of foreign materials and substances
- knock-on effects to ecosystem of impacts on invertebrate populations
- the destruction of places inhabited by plants and animals.

Construction activities can also have an indirect effect on ecological features downstream and off site. The large scale of many linear projects makes it imperative to consider and plan for potential indirect effects.

It is the responsibility of those working on site to ensure that particular species and designated sites are protected.

Site staff are not expected to be ecological experts. However, they are expected to be reasonably aware of potential problems and to seek advice if necessary.

Once the nature conservation body has agreed a construction methodology, it is important to follow it, as inappropriate works or badly managed procedures can also result in prosecution. Requirements may include works methodology, pollution controls, installation of specially designed exclusion fencing, restrictions on the timing of works and specification of certain plant and equipment.

CIRIA C587 *Working with wildlife* (Newton *et al*, 2004a) provides species briefing sheets and toolbox talks covering common and protected species. These are a useful reference if certain species are found on site during construction.

> **What to do where protected species are discovered when the contractor is already on site and works have begun**
>
> Stop work immediately. Seek advice from the nature conservation body on how to proceed. Negotiations may then have to take place to discuss the best way forward.

24.3 VEGETATION CLEARANCE AND LANDSCAPING

The environmental regulator must be consulted and its written permission obtained before herbicides are used in or near controlled waters. (Application form WQM1 can be obtained from the Environment Agency's National Customer Contact Centre.) Only herbicides from an approved list are acceptable and their use must be in accordance with the manufacturer's product label. Staff applying herbicides in or near water must hold a certificate of competence (details available from the National Proficiency Training Council). Extra precautions should be taken in locations where water may be used for the irrigation of crops, watering livestock, fisheries or in fish farming.

During reinstatement, reseeding or landscaping, care should be taken when using fertilisers, as they could enter and pollute watercourses through runoff.

> **Key guidance**
>
> MAFF (1995). PB 2289 *Guidelines for the use of herbicides on weeds in or near watercourses and lakes*, available from Defra, advises on the precautions that should be taken
>
> HSC (1995). L9 *Safe use of pesticides for non-agricultural purposes*
>
> Scottish Executive (2005). *Code of good practice for the prevention of environmental pollution from agricultural activity*
>
> Environment Agency National Customer Contact Centre, tel: 08708 506506.

> **Protecting ecology**
>
> 1. Identify any protected sites or species from a desk-based study, consultation with the nature conservation body or, where available, a review of the environmental statement.
> 2. Commission ecological surveys to identify protected species where necessary.
> 3. Agree a works methodology with the conservation body and ensure all required permissions are in place for each designated area or species encountered.
> 4. Make sure you comply with the agreed methodologies throughout the works.

Appendices

A1	**EIA legislation in relation to linear projects**	215
	A1.1 England and Wales	215
	A1.2 Scotland	216
	A1.3 Northern Ireland	216
A2	**Calculating site runoff rates**	217
	A2.1 Smaller catchments (< 0.5 km^2)	217
	A2.2 Medium-sized catchments (0.5 km^2 to 25 km^2)	218
A3	**Guidance on the optimal timing for carrying out ecological surveys and mitigation**	221
A4	**Internationally, nationally and locally designated sites**	223
	A4.1 Internationally designated sites	223
	A4.2 Nationally designated sites	224
	A4.3 Locally designated sites	224

A1 EIA legislation in relation to linear projects

A1.1 ENGLAND AND WALES

Sector	Regulation	SI no
Planning	Town and Country Planning (Environmental Impact Assessment) (England and Wales) Regulations 1999	1999/293
	Town and Country Planning (Environmental Impact Assessment) (England and Wales) (Amendment) Regulations 2000	2000/2867
Land drainage	Environmental Impact Assessment (Land Drainage Improvement Works) Regulations 1999	1999/1783
Transport	Highways (Assessment of Environmental Effects) Regulations 1999	369/1999
	Transport and Works (Assessment of Environmental Effects) Regulations 2000	2000/3199
	Transport and Works (Assessment of Environmental Effects) Regulations 1998 [England and Wales]	1998/2226
	Transport and Works (Assessment of Environmental Effects) Regulations 1995	1995/1541
Ports and harbours	The Harbour Works (Environmental Impact Assessment) Regulations 1999 [England and Wales, Scotland]	1999/3445
	The Harbour Works (Environmental Impact Assessment) (Amendment) Regulations 2000 [England and Wales, Scotland]	2000/2391
Energy	The Electricity Works (Environmental Impact Assessment) (England and Wales) Regulations 2000	2000/1927
	The Offshore Petroleum Production and Pipe-lines (Assessment of Environmental Effects) Regulations 1999 [England and Wales, Scotland, Northern Ireland]	1999/360
	The Public Gas Transporter Pipe-line Works (Environmental Impact Assessment) Regulations 1999 [England and Wales, Scotland]	1999/1672
	The Pipe-line Works (Environmental Impact Assessment) Regulations 2000 [England and Wales, Scotland, Northern Ireland]	2000/1928
Water abstraction	Water Resources (Environmental Impact Assessment) (England and Wales) Regulations 2003	2003/164

A1.2 SCOTLAND

Sector	Regulation	SI or SSI no
Planning	Environmental Impact Assessment (Scotland) Regulations 1999	SSI 1999/1
	Environmental Impact Assessment (Scotland) (Amendment) Regulations 2002	SSI 2002/324
Land drainage	Environmental Impact Assessment (Scotland) Regulations 1999 – Part IV Drainage Works	SSI 1999/1
Transport	Environmental Impact Assessment (Scotland) Regulations 1999 – Part III Roads	SSI 1999/1
Ports and harbours	The Harbour Works (Environmental Impact Assessment) Regulations 1999 [*England and Wales, Scotland*]	SI 1999/3445
	The Harbour Works (Environmental Impact Assessment) (Amendment) Regulations 2000 [*England and Wales, Scotland*]	SI 2000/2391
Energy	Environmental Impact Assessment (Scotland) Regulations 1988 – Part III Electricity	SSI 1988/1221
	The Offshore Petroleum Production and Pipe-lines (Assessment of Environmental Effects) Regulations 1999 [*England and Wales, Scotland, Northern Ireland*]	SI 1999/360
	The Public Gas Transporter Pipe-line Works (Environmental Impact Assessment) Regulations 1999 [*England and Wales, Scotland*]	SI 1999/1672
	The Pipe-line Works (Environmental Impact Assessment) Regulations 2000 [*England and Wales, Scotland, Northern Ireland*]	SI 2000/1982

A1.3 NORTHERN IRELAND

Sector	Regulation	SI or SR no
Planning	Planning (Environmental Impact Assessment) Regulations (Northern Ireland) 1999	SR 1999/73
Land drainage	Drainage (Environmental Impact Assessment) Regulations (Northern Ireland) 2001	SR 2001/394
Transport	Roads (Environmental Impact Assessment) Regulations (Northern Ireland) 1999	SR 1999/98
Ports and harbours	Harbour Works (Assessment of Environmental Effects) Regulations (Northern Ireland) 1990	SR 1990/181
Energy	The Offshore Petroleum Production and Pipe-lines (Assessment of Environmental Effects) Regulations 1999 [*England and Wales, Scotland, Northern Ireland*]	SI 1999/360
	The Pipe-line Works (Environmental Impact Assessment) Regulations 2000 [*England and Wales, Scotland, Northern Ireland*]	SI 2000/1928

A2 Calculating site runoff rates

A2.1 SMALLER CATCHMENTS (< 0.5 KM²)

This method uses annual rainfall and soil type data to calculate the mean annual flood for catchments smaller than 0.5 km². Flood flows can be calculated using Table A2.1, which provides estimates of mean annual flood in litres/second/hectare.

Table A2.1 *Mean annual flood peak flow for catchments < 50 ha (l/s/ha)*

Soil type	Annual rainfall (mm)					
	< 600	600–800	800–1200	1200–1600	1600–3200	> 3200
1	0.3	0.4	0.6	0.9	1.7	2.4
2	1.4	1.8	2.8	4.1	7.7	10.8
3	2.6	3.4	5.2	7.7	14.4	20.1
4	3.3	4.4	6.7	9.9	18.6	26.0
5	4.2	5.5	8.4	12.4	23.3	32.7

A2.1.1 Worked example 1

This example is for a site in Carmarthenshire. Text in italics originates in Chapter 18.

Parameter	Derivation	Value
Rainfall zone	*Annual rainfall can be estimated from Figure 18.1, or known site-specific values can be used.* The site-specific value is not known so an estimate is used from Figure 18.1.	1200–1600 mm
Soil type	*Soils are divided into the five classes shown in Table 18.3.* The soil is a sandy clay. It was decided that runoff potential was "Moderate".	3
Peak flow per hectare	*Flood flows should be estimated from Table 18.2.* See Table A2.1 above.	7.7 l/s/ha
Mean annual flood	*Multiply this flood flow in litres/second/hectare by the catchment area (in hectares).* The catchment area here is 0.26 km². Mean annual flood for the catchment = 7.7 l/s/ha × 26 ha.	200.2 l/s
10-year return period flood	*The mean annual flood can be multiplied by a factor for range of return periods (Table 18.4).* Ie a site peak flow for a 10-year return period = 200.2 l/s × 1.48.	296.3 l/s

The 10-year return period flood for the catchment is approximately 296.3 l/s.

A2.1.2 Worked example 2

This example is for a site in Norfolk. Text in italics originates in Chapter 18.

Parameter	Derivation	Value
Rainfall zone	*Annual rainfall can be estimated from Figure 18.1, or known site-specific values can be used.* The site-specific value annual average rainfall is known to be around 632 mm.	632 mm
Soil type	*Soils are divided into the five classes shown in Table 18.3.* The soil is very permeable and loamy. It was decided that runoff potential was "Very low".	1
Peak flow per hectare	*Flood flows should be estimated from Table 18.2.* See Table A2.1 above.	0.4 l/s/ha
Mean annual flood	*Multiply this flood flow in litres/second/hectare by the catchment area (in hectares).* The catchment area here is only 0.0048 km² as it is in an urban area. Mean annual flood for the catchment = 0.4 l/s/ha × 0.48 ha.	0.19 l/s
Mean annual flood for a 50-year return period	*The mean annual flood can be multiplied by a factor for range of return periods (Table 18.4).* Ie a site peak flow for a 50-year return period = 0.19 l/s × 2.83.	0.54 l/s

The mean annual flood for the catchment = 0.19 l/s.

(The mean annual flood with 10-return year period would be 0.28 l/s.)

A2.2 MEDIUM-SIZED CATCHMENTS (0.5 KM² TO 25 KM²)

Calculation of runoff rates using Report 124 *Flood estimation for small catchments* **(IH, 1994).**

This method estimates mean annual flood for greenfield conditions using parameters derived for the *Flood studies report* (IH, 1974).

The estimation equation is:

Mean annual flood = $0.00108 \times AREA^{0.89} \times SAAR^{1.17} \times SOIL^{2.17}$

AREA = catchment area (km²)

SAAR = standard average annual rainfall 1961–1990 (mm)

SOIL = soil index = $(0.15 S_1 + 0.3 S_2 + 0.4 S_3 + 0.45 S_4 + 0.5 S_5)/(S_1 + S_2 + S_3 + S_4 + S_5)$

S_1, S_2 etc are soil characteristics derived from mapped data.

A2.2.1 Worked example 3

This example is for a site in Lancashire where a natural peak runoff rate from a catchment was required to design a temporary crossing.

Parameter	Derivation	Value
AREA	The area of the catchment is 73 ha = 0.73 km² Note: the recommendation (Defra/EA, 2004) for sites less than 50 ha is to use 50 ha in the equation and reduce the result pro rata for catchment area.	0.73 km²
SAAR	SAAR is the catchment annual average rainfall	980 mm
SOIL	The soil is clayey and relatively thin. It was decided that runoff potential was "high", ie soil type 4. This gives a value for SOIL of 0.45	0.45

Using these values, the mean annual flood = $0.00108 \times (0.73)^{0.89} \times (980)^{1.17} \times (0.45)^{2.17}$

ie mean annual flood = 0.456 m³/s, or 456 l/s.

Flood estimates for other return periods are derived from scaling factors in Flood Studies Supplementary Report 14 *Review of regional growth curves* (IH, 1983), reproduced in Table 18.5. An extract from Table 18.5, for sites in north-west England (used for this example), is shown in the following table with their resultant peak flow estimates for return periods of five, 10 and 50 years.

Return period (years)	Scaling factor	Peak flow m³/s	Peak flow l/s
5	1.19	0.543	543
10	1.38	0.629	629
50	1.85	0.844	844

A3 Guidance on the optimal timing for carrying out ecological surveys and mitigation

Protected species	Jan	Feb	Mar	Apr	May	Jun	Jul	Aug	Sep	Oct	Nov	Dec
Habitats and vegetation	Recommended time to survey for mosses and lichens only			Recommended time to undertake Phase 1 habitat survey						Recommended time to survey for mosses and lichens only		
	Best time for planting and translocation			Planting and translocation of majority of species not recommended						Best time for planting and translocation		
Birds		Bird nesting season – no work if nesting birds found						Extended for some species				
	Vegetation can be cleared but stop immediately if nesting birds found									Vegetation can be cleared but stop immediately if nesting birds found		
Badgers	No work close to setts						Badger exclusion licensing season					No work close to setts
	Best time for field surveys					Surveys may still be carried out, but they will not be as effective				Best time for field surveys		
	Construction of artificial setts		Territorial bait surveys can be carried out									Construction of artificial setts
Bats	No work in hibernation roosts					No work in breeding roosts						No work in hibernation roosts
Dormice			If dormice seem to be present stop work, report your findings and put a protection plan into operation									
	Hibernation period – good time to survey for nests								Hibernation period – good time to survey for nests			
Red squirrels			Very timid and extremely rare – if spotted, stop work and report the sighting									
		Breeding season; particularly vulnerable										
Water voles	Avoid all works in water vole habitat			Work in water vole habitat possible	Avoid all works in water vole habitat			Work in water vole habitat possible		Avoid all works in water vole habitat		
				Best time to survey								
Reptiles	Hibernation period – avoid work			Best time for capture				Best time for capture		Hibernation period – avoid work		
Great crested newts	No work in inhabited ponds			Best time for licensing and surveying							No work in inhabited ponds	
				Best time for terrestrial work								
Natterjack toads	Do not disturb when hibernating					Do not work in ponds containing natterjack toads or spawn; do not move them without a licence				Do not disturb when hibernating		
				Best time for survey in ponds		Pond survey OK, but numbers will be lower		Suitable time to survey for adults on land				
Fish					Be aware that this is the time for fish spawning and salmon runs – seek advice							

Figure A3.1 *Wildlife year planner*

Work restricted   Recommended survey time   Recommended time to work

A4 Internationally, nationally and locally designated sites

A4.1 INTERNATIONALLY DESIGNATED SITES

Designation	Supported by
Ramsar site Ramsar sites are designated as SPAs and protected as SSSIs in Britain and as ASSIs in Northern Ireland	Convention on Wetlands of International Importance Especially Water Fowl Habitat 1971 (Ramsar Convention)
Biosphere reserve Receives statutory protection as an NNR	UNESCO Man and the Biosphere Programme 1970
Biogenic reserve Receives statutory protection as an SSSI or NNR	Berne Convention 1979
World heritage site Among these are features such as Hadrian's Wall, Stonehenge and the New Lanark Industrial Landscape	UNESCO Convention for the Protection of World Culture and Natural Heritage 1972
European site	EC Habitats Directive 1992 and: • Special areas of conservation – UK Habitats Regulations 1994 • Special protection areas – EC Wild Birds Directive 1979 • Sites of community importance
Candidate or potential European site	EC Habitats Directive 1992 and: • Candidate SACs – UK Habitats Regulations 1994 • Potential SPAs – EC Wild Birds Directive 1979
European diploma site	Council of Europe
Site hosting habitats or species of (European) Community interest	Annexes 1 and 2 of the Habitats Directive 1992
Site hosting significant species populations under the Bonn Convention	Convention on the Conservation of Migratory Species of Wild Animals, 1979
Site hosting significant populations under the Berne Convention	Convention on the Conservation of European Wildlife and Natural Habitats, 1979

A4.2 NATIONALLY DESIGNATED SITES

Designation	Supported by
Sites of special scientific interest (SSSIs), areas of special scientific interest (ASSIs) (NI)	National Parks and Access to the Countryside Act (NP&AC) 1949; Wildlife and Countryside Act 1981; CRoW Act 2000; Environment (NI) Order 2003
Nature conservation order	Wildlife and Countryside Act 1981
Special nature conservation order	Habitat Regulations 1994
Marine nature reserve	Wildlife and Countryside Act 1981, Nature Conservation and Amenity Lands (NCAL) (NI) Order 1985
Area of special protection for birds	Wildlife and Countryside Act 1981
Bird sanctuary	Protection of Birds Act 1954
National park	NP&AC Act 1949 (as amended), NCAL (NI) Order 1985
Area of outstanding natural beauty, national scenic area (in Scotland)	NP&AC Act 1949 (as amended), NCAL (NI) Order 1985
Environmentally sensitive area	Agriculture Act 1986 (as amended), Agriculture (Environmental Areas) (NI) Order 1987
Natural heritage area	Natural Heritage (Scotland) Act 1991
Limestone pavement order	Wildlife and Countryside Act 1981
Nature conservation review site	Listed in the Nature Conservation Review (NCR)
Geological conservation review (GCR) site	
Sites hosting *Red data book* species	
Sites hosting species not covered by Berne Convention but in Schedules 1,5 and 8 of the WCA 1981	

A4.3 LOCALLY DESIGNATED SITES

Designation	Supported by
Local nature reserve	National Parks and Access to the Countryside Act 1949 (as amended), NCAL (NI) Order 1985
Site of importance for nature conservation (SINC) Site of nature conservation importance (SNCI) County wildlife site or similar	Usually confirmed by the LPA in conjunction with the local wildlife trust and listed within attendant policies in the respective local plan
Regionally important geological site (RIG)	
Important "inventory" site (eg inventory of ancient semi-natural woodland and grassland	Usually kept by the statutory nature conservation organisation (SNCO)
Site other than one described above that contains biodiversity action plan (BAP) priority habitats or species	Listed in the local biodiversity action plan
Other natural or semi-natural site of significant biodiversity importance, not referred to above, such as site relevant to local BAP or natural area objectives	Possibly listed in the local biodiversity action plan
Site not in the above categories, but with some biodiversity or earth heritage interest	Could be any site (eg a notified hedgerow)

References and bibliography

Aspinwall & Co (1994). *A framework for assessing the impact of contaminated land on groundwater and surface water*. CLR1, Department of the Environment, London, vols 1 and 2

Auckland Regional Council (2004). "The use of flocculants and coagulants to aid the settlement of suspended sediment in earthworks runoff: trials, methodology and design". Technical Publication 227 (draft), Auckland Regional Council, Auckland, <www.arc.govt.nz/library/I35022_2.pdf>

Australian Capital Territory (1998). *Erosion and sediment control during land development*. Environment ACT, Canberra, <www.environment.act.gov.au>

Barr, D; Finnamore, J R; Bardos, R P; Weeks, J M and Nathanail, C P (2002). *Biological methods for assessment and remediation of contaminated land – case studies*. C575, CIRIA, London

Barr, D; Bardos, R P and Nathanail, C P (2003). *Non-biological methods for assessment and remediation of contaminated land – case studies*. C588, CIRIA, London

Beckstrand, D; Smith, P; Gaffney, F; Dickerson, D; Walowsky, D; Karimipour, S and Lake, D W (2004). *Draft New York standards and specifications for erosion and sediment control*. New York State Department of Environmental Conservation, Albany, NY, <www.dec.state.ny.us/website/dow/toolbox/escstandards>

Bettess, R (1996). *Infiltration drainage – manual of good practice*. R156, CIRIA, London

Boland, M P; Klinck, BA; Robins, N S; Stuart, M E and Whitehead, E J (1999). *Guidelines and protocols for investigations to assess site specific groundwater vulnerability*. P2/042/01, Environment Agency, Bristol, <http://publications.environment-agency.gov.uk/pdf/SPRP2-042-01-e-p.pdf>

BRE (1991). *Soakaway design* (revised 2003). Digest 365, BRE Books, Garston, 8 pp (downloadable PDF), ISBN 1-86081-604-5

British Waterways (2005). *Code of practice for works affecting British Waterways*. British Waterways, Watford, <www.britishwaterways.co.uk/images/COP_2005.pdf>

Budd, S; John, S; Simm, J and Wilkinson, M W (2003). *Coastal and marine environmental site guide*. C584, CIRIA, London

Centre for Aquatic Plant Management (1995). *Guidelines for the use of herbicides on weeds in or near watercourses and lakes*. PB 2289, Ministry of Agriculture, Fisheries and Food, London. Available free from Defra

Chant-Hall, G; Charles, P and Connolly, S (2005a). *Environmental good practice on site (second edition)*. C650, CIRIA, London

Chant-Hall, G; Charles, P and Connolly, S (2005b). *Environmental good practice on site – pocket book*. C651, CIRIA, London

CIRIA (1995). *A client's guide to greener construction*. SP120, CIRIA, London

CIRIA (1998a). *Septic tank systems: a regulator's guide*. SP144BT, CIRIA, London

CIRIA (1998b). *Septic tank systems, 2: options*. SP144L2, CIRIA, London

CIRIA (1998c). *Septic tank systems, 3: design and installation*. SP144L3, CIRIA, London

CIRIA (2005). *Remedial treatment for contaminated land, Volumes I–XII*. Set of 12 books formerly issued as SP101 to SP112. SP164, CIRIA, London

CIRIA and Environment Agency (1996). *Building a cleaner future*. SP141V, CIRIA, London

City of Houston, Harris County and Harris County Flood District (2001). *Storm water management handbook for construction activities*. Storm Water Management Joint Task Force, Houston, <www.cleanwaterclearchoice.org>, accessed 7 Jun 2005

Clarke, B G and Lawson, C (2005). *Slurry: management and disposal of semi-liquid spoil*. Pipe Jacking Association, London

Construction Confederation Environmental Forum (2005). *Environmental induction video and toolbox talks*. Construction Confederation, London, <www.thecc.org.uk/>

Construction Products Association (2005). *Construction products industry key performance indicators. 2005 handbook* (updated annually). Construction Products Association, London, <www.constprod.org.uk/download/network/CPI%20KPI%202005%20-%20Handbook.pdf>

Coppin, N J and Richards, I G (1990). *Use of vegetation in civil engineering*. B10, CIRIA, London, and Butterworths, London

Dee, T and Sivil, D (2001). *Selecting package wastewater treatment plants*. PR72, CIRIA, London

Department for Environment, Food and Rural Affairs (2005). *Guidance on mixing hazardous waste. Hazardous Waste Regulations*. Defra, London, <www.defra.gov.uk/environment/waste/special/pdf/hwrmixing-guide.pdf>

Department for Environment, Food and Rural Affairs and Environment Agency (2004). *Preliminary rainfall runoff management for developments*. Technical report W5-074A/TR/1 Revision B, Defra, London and EA, Bristol

Department for Environment, Transport and the Regions (1999). *Planning requirement in respect of the use of non-mains sewerage incorporating septic tanks in new development*. Planning Circular 03/99, Stationery Office, London, <www.odpm.gov.uk/>

Department for Transport, Local Government and the Regions (2001). *Development and flood risk*. Planning Policy Guidance Note 25, DTLR, London, <www.odpm.gov.uk/index.asp?id=1144113>

Department of the Environment (nd). "Industry Profiles". A series of guidance documents on possible contamination from various industries. DoE, London, <www.environment-agency.gov.uk/subjects/landquality/113813/1166435/?lang=_e>

Department of the Environment (1995). *General Development Order consolidation 1995*. DOE Circular 9/95, HMSO, London

Department of the Environment (1996). *Waste management – the duty of care: a code of practice*. HMSO, London, <www.defra.gov.uk/environment/waste/management/doc/pdf/waste_man_duty_code.pdf>

Department of the Environment (Northern Ireland) (1999). *The Groundwater Regulations (Northern Ireland) 1998. DOE (NI) Guidance Note 2. Disposal of List I and II substances to land: general guidance on compliance*. DoE (NI), Belfast, <www.ehsni.gov.uk/pubs/publications/GroundwaterGuidanceNote2v1-1.pdf>

Department of Trade and Industry (2004). *Site waste management plans. Guidance for construction contractors and clients. Voluntary code of practice*. DTI, London <www.dti.gov.uk/construction/sustain/site_waste_management.pdf>

Environment Agency (1999). *Waterway bank protection: a guide to erosion assessment and management*. R&D Publication No 11, EA, Bristol

Environment Agency (2001a). *Conserving water in buildings*. Series of fact cards. EA, Bristol, <www.environment-agency.gov.uk/subjects/waterres/>

Environment Agency (2001b). *Piling into contaminated sites*. EA, Bristol

Environment Agency (2001c). *Piling and penetrative ground improvement methods on land affected by contamination – guidance on pollution prevention*. NC/99/73, EA, Bristol

Environment Agency (2004a). *The Water Act 2003. Modernising the regulation of water resources*. EA, Bristol, <www.environment-agency.gov.uk>

Environment Agency (2004b). *Best practice techniques for service crossings of watercourses*. Consultation draft. EA, Bristol, <www.environment-agency.gov.uk>

Environment Agency (2005a). *Groundwater protection: policy and practice*. EA, Bristol

Environment Agency (2005b). *State of the environment*. EA, Bristol, <www.environment-agency.gov.uk>

Environment Agency, SEPA and EHS (2000). *Works in, near or liable to affect water*. PPG5, EA, Bristol; SEPA, Edinburgh; EHS, Belfast, <www.netregs.gov.uk/commondata/acrobat/ppg05.pdf>

Environment Agency, SEPA and EHS (2001). *High pressure water and steam cleaners*. PPG13, EA, Bristol; SEPA, Edinburgh; EHS, Belfast, <www.sepa.org.uk/pdf/guidance/ppg/ppg13.pdf>

Environment Agency, SEPA and EHS (2004a). *Pollution incident response planning*. PPG21, EA, Bristol; SEPA, Edinburgh; EHS, Belfast, <www.sepa.org.uk/pdf/guidance/ppg/ppg21.pdf>

Environment Agency, SEPA and EHS (2004b). *Disposal of sewage where no main drainage is available*. PPG4, EA, Bristol; SEPA, Edinburgh; EHS, Belfast, <www.sepa.org.uk/pdf/guidance/ppg/ppg04.pdf>

Environment Agency, SEPA and EHS (2004c). *Above ground oil storage tanks*. PPG2, EA, Bristol; SEPA, Edinburgh; EHS, Belfast, <www.sepa.org.uk/pdf/guidance/ppg/ppg02.pdf>

Environment Agency, SEPA and EHS (2004d). *Refuelling facilities*. PPG7, EA, Bristol; SEPA, Edinburgh; EHS, Belfast, <www.sepa.org.uk/pdf/guidance/ppg/ppg07.pdf>

Environment Agency, SEPA and EHS (2004e). *Safe storage and disposal of used oils*. PPG8, EA, Bristol; SEPA, Edinburgh; EHS, Belfast, <www.sepa.org.uk/pdf/guidance/ppg/ppg08.pdf>

Environment Agency, SEPA and EHS (2004f). *Storage and handling of drums and intermediate bulk containers*. PPG26, EA, Bristol; SEPA, Edinburgh; EHS, Belfast, <www.sepa.org.uk/pdf/guidance/ppg/ppg26.pdf>

Environment Agency, SEPA and EHS (2004g). *Getting your site right. Industrial and commercial pollution prevention*. EA, Bristol, <www.environment-agency.gov.uk/commondata/acrobat/pp_pays_booklet_e_1212832.pdf>

Environment Agency, SEPA and EHS (2004h). *Use and design of oil separators in surface water drainage systems*. PPG3, EA, Bristol; SEPA, Edinburgh; EHS, Belfast, <www.sepa.org.uk/pdf/guidance/ppg/ppg03.pdf>

Environment Agency, SEPA, EHS and CIRIA (2000a). *Masonry bunds for oil storage tanks*. EA, Bristol; SEPA, Edinburgh; EHS, Belfast; CIRIA, London <www.sepa.org.uk/pdf/guidance/ppg/masonrybunds.pdf>,

Environment Agency, SEPA, EHS and CIRIA (2000b). *Concrete bunds for oil storage tanks*. EA, Bristol; SEPA, Edinburgh; EHS, Belfast; CIRIA, London, <www.sepa.org.uk/pdf/guidance/ppg/concretebunds.pdf>

Environment and Heritage Service (2000). Water (Northern Ireland) Order 1999. Guide for vehicle wash operators. Treatment and disposal options. EHS, Belfast, <www.ehsni.gov.uk/pubs/publications/Vehicle_Wash_Operators.pdf>

Environment and Heritage Service (2001). *Policy and practice for the protection of groundwater in Northern Ireland*. EHS, Belfast, <www.ehsni.gov.uk/pubs/publications/Policy_and_Practice_for_the_Protection_of_Groundwater_in_Northern_Ireland.pdf>

Envirowise (2005). *Cost-effective water saving devices and practices – for commercial sites*. GG067, Envirowise, London

Evans, D; Jefferis, S A; Thomas, A O and Cui, S (2001). *Remedial processes for contaminated land – principles and practice*. C549, CIRIA, London

Fisher, K and Ramsbottom, D (2001). *River diversions: a design guide*. Thomas Telford, London

Gaffney, F B and Lake, D W (2004). "Revised universal soil loss equation". In: D Beckstrand *et al*, *Draft New York standards and specifications for erosion and sediment control*. New York State Department of Environmental Conservation, Albany, NY, <www.dec.state.ny.us/website/dow/toolbox/escstandards/3rusle.pdf>

Gannon, J (1999). "A review of sediment control measures". *Erosion control*, <www.clearcreeksystems.com/sediment/erosion.htm>

Godfrey, P S (1996). *Control of risk: a guide to the systematic management of risk from construction*. SP125, CIRIA, London

Gyasi-Agyei, Y; Sibley, J and Ashwath, N (2001). "Quantitative evaluation of strategies for erosion control on a railway embankment batter". *Hydrological processes*, 31 Nov, vol 15, pp 3249–3268

Hall, M J; Hockin, D L and Ellis, J B (1993). *Design of flood storage reservoirs*. B14, CIRIA, London

Harbor, J (1999). "Engineering geomorphology at the cutting edge of land disturbance: erosion and sediment control on construction sites". *Geomorphology*, vol 31, pp 247–263

Health and Safety Commission (1995). *Safe use of pesticides for non-agricultural purposes*. L9, HSE Books, Ipswich

Highways Agency (1998). *Design manual for roads and bridges* (DMRB). Stationery Office, London, <www.archive2.official-documents.co.uk/document/deps/ha/dmrb/>

Highways Agency (nd). "Environmental design and management". In: *Design manual for roads and bridges*, vols 10 and 10a, Stationery Office, London

Highways Agency (nd). "Environmental assessment". In: *Design manual for roads and bridges*, vol 11, Stationery Office, London

Highways Agency (2001). "Vegetative treatment systems for highway runoff". In: *Design manual for roads and bridges*, vol 4, sec 2, pt 1, HA 103/01, Stationery Office, London

Highways Agency (2004). "Drainage of runoff from natural catchments". In: *Design manual for roads and bridges*, vol 4, sec 2, pt 1, HA 106/04, HA 103/01, Stationery Office, London

Hill, K (2005). "Silt pollution of waterways from construction sites: a problem the industry must learn to address". *EDIE*, Apr, <www.edie.net>, accessed 30 Jun 2005

Holden, J M W; Jones, M A; Mirales-Wilhelm, F and White, C (1998). *Hydraulic measures for the control and treatment of groundwater pollution*. R186, CIRIA, London

HR Wallingford and Institute of Hydrology (1981). *The Wallingford Procedure: design and analysis of urban storm drainage*. Three-volume set. HR Wallingford, Wallingford

Institute of Hydrology (1974). *Flood studies report*. Institute of Hydrology, Wallingford

Institute of Hydrology (1983). *Review of regional growth curves*. Flood Studies Supplementary Report 14, Institute of Hydrology, Wallingford

Institute of Hydrology (1994). *Flood estimation for small catchments*. Report 124, Institute of Hydrology, Wallingford

Institute of Hydrology (1999). *Flood estimation handbook*. Institute of Hydrology, Wallingford

Jurries, D (1999). *Flocculation of construction site runoff*. Oregon Department of Environmental Quality, Portland, <www.deq.state.or.us/nwr>, accessed 5 May 2005

Laidler, D W; Bryce, A J and Wilbourn, P (2002). *Brownfields – managing the development of previously developed land. A client's guide*. C578, CIRIA, London

Leeds-Harrison, P B *et al* (1996). *Buffer zones in headwater catchments*. Annex Six: The use of buffer zones to protect water courses from sediment and sediment bound contaminants. MAFF/English Nature Buffer Zone Project CSA 2285, Cranfield University, Cranfield

London Underground (nd). *Guidance on drainage design, installation, maintenance and removals*. London Underground, London

Marsland, P A and Carey, M A (1999). *Method for the derivation of remedial targets for soil and groundwater to protect water resources*. R&D Publication 20, Environment Agency, Bristol, <http://publications.environment-agency.gov.uk/pdf/SR-DPUB20-e-e.pdf>

Martin, P; Turner, B; Waddington, K; Pratt, C; Campbell, N; Payne, J and Reed, B (2000a). *Sustainable urban drainage systems – design manual for Scotland and Northern Ireland*. C521, CIRIA, London

Martin, P; Turner, B; Waddington, K; Dell, J; Pratt, C; Campbell, N; Payne, J and Reed, B (2000b). *Sustainable urban drainage systems – design manual for England and Wales*. C522, CIRIA, London

Mason, P A; Amies, H J; Sangarapillai, G and Rose, G (1997). *Construction of bunds for oil storage tanks*. R163, CIRIA, London

Masters-Williams, H; Heap, A; Kitts, H; Greenshaw, L; Davis, S; Fisher, P; Hendrie, M and Owens, D (2001). *Control of water pollution from construction sites. Guidance for consultants and contractors*. C532, CIRIA, London

McLaughlin, R A (2002). *Measures to reduce erosion and turbidity in construction site runoff*. FHWA/NC/2002-023, North Carolina State University, Raleigh <www.ncdot.org>, accessed 5 May 2005

McNeill, A (1996). "Road construction and river pollution in southwest Scotland". *J CIWEM*, 10 Jun

Ministry of Agriculture, Fisheries and Food (1995). *Guidelines for the use of herbicides on weeds in or near watercourses and lakes*. MAFF, London

Morledge, R and Jackson, F (2001). "Reducing environmental pollution caused by construction plant". *Environmental Management and Health*, vol 12, pp 191–206

Morris, M W and Simm, J D (eds) (2000). *Construction risk in river and estuary engineering. A guidance manual*. Thomas Telford, London

Murnane, E; Heap, A; Grimes, J; Rawlinson, J; Williams, I and Forrester, L (2002). *Control of water pollution from construction sites – guide to good practice*. SP156, CIRIA, London

Network Rail (nd). *Design manual for rail drainage systems*. Network Rail, London

Newton, J; Williams, C; Nicholson, B; Venables, R; Willetts, R and Moser, B (2004a). *Working with wildlife. A resource and training pack for the construction industry*. C587, CIRIA, London

Newton, J; Williams, C; Nicholson, B; Venables, R; Willetts, R and Moser, B (2004b). *Working with wildlife pocket book*. C613, CIRIA, London

North Carolina State University (nd). *Woodsong – sediment research*. North Carolina State University, Raleigh, <www.ces.ncsu.edu/depts/design/research/woodsong/sediment.html>, accessed 7 Jun 2005

Office of the Deputy Prime Minister; Department for Transport; Department for Environment, Food and Rural Affairs; and Office of Government Commerce (2002). *Green public private partnerships*. ODPM, DfT, Defra and OGC, Norwich, <www.hm-treasury.gov.uk/media/851/A5/PPP_GreenPublicPrivatePart.pdf>

Oregon Department of Environmental Quality (2004). *Best management practices for storm water discharges associated with construction activities*. Oregon Department of Environmental Quality, Portland, <www.deq.state.or.us/WQConstructionBMPs.pdf>

Osborne, M and Montague, K (2005). *The potential for water pollution from railways*. C643, CIRIA, London

Palmer, J (2000). Ch 18, "Dewatering". In: Waterway Recovery Group, *Practical restoration handbook*. Inland Waterways Association, Rickmansworth, <www.wrg.org.uk/prh/dewatering.pdf>

Pipe Jacking Association (1995a). *An introduction to pipe jacking and microtunnelling design*. Pipe Jacking Association, London

Pipe Jacking Association (1995b). *Guide to best practice for the installation of pipe jacks and microtunnels*. Pipe Jacking Association, London

Pipe Jacking Association (in prep). *Management of process arisings from tunnels and other earthworks: a guide to regulatory compliance*. Pipe Jacking Association, London

Pipeline & Plant Construction Group Environmental Forum (2003). Internal document. PCEF

Postle, M and Vernon, J (2002). *Guidance on the costing of environmental pollution from construction*. C565, CIRIA, London

Preene, M; Roberts, T O L; Powrie, W and Dyer, M R (2000). *Groundwater control – design and practice*. C515, CIRIA, London

Privett, K D; Matthew, S and Hodges, R A (1996). *Barriers, liners and cover systems for containment and control of land contamination*. SP124, CIRIA, London

Prosser (1992). *Design of low-lift pumping stations – with particular application to pumping wastewater*. Report 121, CIRIA, London

Railtrack (2002). *Company standards. Railtrack contract requirements – environment*. RT/LS/S/015, Railtrack, London, <www.larif.org.uk/Network%20Rail%20Code.pdf>

River Restoration Centre (2002). *Manual of river restoration techniques: 2002 update*. River Restoration Centre, Silsoe, <www.therrc.co.uk/manual.php>

Rudland, D J and Jackson, S D (2004). *Selection of remedial treatments for contaminated land. A guide to good practice*. C622, CIRIA, London

Rudland, D J; Lancefield, R M and Mayell, P N (2001). *Contaminated land risk assessment. A guide to good practice*. C552, CIRIA, London

Scottish Environmental Protection Agency (2002). *The future for Scotland's waters, guiding principles on the technical requirements of the Water Framework Directive*. SEPA, Edinburgh

Scottish Environmental Protection Agency (2003). *Groundwater protection policy for Scotland*. EP 019, SEPA, Edinburgh, <www.sepa.org.uk/pdf/policies/19.pdf>

Scottish Executive (2000). *River crossings and migratory fish: design guidance*. Scottish Executive Development Department, Edinburgh, <www.scottishexecutive.gov.uk/consultations/transport/rcmf-03.asp>

Scottish Executive (2005). *Prevention of environmental pollution from agricultural activity. Code of good practice. Dos and don'ts guide*. Scottish Executive, Edinburgh, <www.scotland.gov.uk/library5/environment/pepf.pdf>

Scottish Water (2005). "10 Top Tips". Ten guidance leaflets for business. Scottish Water, Dunfermline, <www.scottishwater.co.uk>

Sealey, B J; Hill, G J and Phillips, P S (2001). "Review of strategy for recycling and reuse of waste materials, recycling and reuse of waste materials". In: R K Dhir, M Limbachiya, M McCarthy (eds), *Proc int symp, Dundee University, 19–20 Mar*. Thomas Telford, London, pp 325–335

Steeds, J E; Shepherd, E and Barry, D L (1996). *A guide for safe working on contaminated sites*. R132, CIRIA, London

Teekaram, A; Sterne, S; Abel, B and Elliott, C (2002). *Above-ground proprietary prefabricated oil storage tank systems*. C535, CIRIA, London

Uren, S and Griffiths, E (2000). *Environmental management in construction*. C533, CIRIA, London

Van Bohemen, H D and Janssen van de Laak, W D (2003). "The influence of road infrastructure and traffic on soil, water and air quality". *Environmental management*, no 31, pp 50–68

Venables, R; Newton, J; Westaway, N; Venables, J; Castle, P; Neale, B; Short, D; McKenzie, J; Leach, A; Housego, D; Chapman, J and Peirson-Hills, A (2000a). *Environmental handbook for building and civil engineering projects. Part 1 Design and specification*. C512, CIRIA, London

Venables *et al* (2000b). *Environmental handbook for building and civil engineering projects. Part 2 Construction*. C528, CIRIA, London

Venables *et al* (2000c). *Environmental handbook for building and civil engineering projects. Part 3 Demolition and site clearance*. C529, CIRIA, London

Vik, E A; Sverdrup, L; Kelley, A; Storhaug, R; Beitnes, A; Boge, K; Grepstad, G K and Tveiten, V (2000). "Experiences from environmental risk management of chemical grouting agents used during construction of the Romeriksporten tunnel". *Tunnelling and Underground Space Technology*, vol 15, pp 369–378

Westcott, F J; Lean, C M B and Cunningham, M L (2001). *Piling and penetrative ground improvement methods on land affected by contamination: guidance on pollution prevention*. NC/99/73, Environment Agency, Bristol, <www.environment-agency.gov.uk/commondata/acrobat/nc_99_73_piling_pdf>

LEGISLATION

European Directives

Council Directive 76/464/EEC of 4 May 1976 on pollution caused by certain dangerous substances discharged into the aquatic environment of the community ("the Dangerous Substances Directive"). *Official Journal* L129 P0023–0029

Council Directive 80/68/EEC of 17 Dec 1979 on the protection of groundwater against pollution caused by certain dangerous products ("the Groundwater Directive"). *Official Journal* L020, P0043–0048

Council Directive 85/337/EEC of 27 Jun 1985 on the assessment of the effects of certain public and private projects on the environment ("the Environmental Impact Assessment Directive"). *Official Journal* L175, P0040–0048. Amended by 97/11/EC and 2003/35/EC

Council Directive 92/43/EEC of 21 May 1992 on the conservation of natural habitats and of wild fauna and flora ("the Habitats Directive"). *Official Journal* L206 P0029–0030

Council Directive 2000/60/EC of the European Parliament and of the Council of 23 Oct 2000 establishing a framework for Community action in the field of water policy ("the Water Framework Directive"). *Official Journal* L327 P0001–0073

UK Acts

Channel Tunnel Rail Link Act 1996, c.61

Coast Protection Act 1949, c.74

Control of Pollution Act 1974, c.40 (as amended) [Scotland]

Electricity Act 1989, c.29 (as amended)

Environmental Act 1995, c.25

Environment Protection Act 1990, c.43

Food and Environment Protection Act 1985, c.48

Harbours Act 1964, c.40

Highways Act 1980, c.66

Land Drainage Act 1991, c.59

Planning and Compulsory Purchase Act 2004, c.5

Roads (Scotland) Act 1984, c.54

Telecommunications Act 1984, c.12

Town and Country Planning Act 1990, c.8 (as amended)

Transport and Works Act 1992, c.42

Water Act 2003, c.37

Water Environment and Water Services (Scotland) Act 2003, asp3

Water Industry Act 1991, c.56

Water Industry (Scotland) Act 2002, asp3

Water Resources Act 1991, c.57

Wildlife and Countryside Act 1981, c.69 (as amended)

UK Regulations and Orders

Building Regulations 2000, SI 2000/2531

Conservation (Natural Habitats, &c) Regulations 1994, SI 1994/2716

Conservation (Nature Habitats, etc) Regulations (Northern Ireland) 1995, SR 1995/380

Construction (Design and Management) Regulations 1994, SI 1994/3140

Controlled Waste (Duty of Care) Regulations (Northern Ireland) 2002, SR 2002/271

Control of Pesticides Regulations 1986, SI 1986/1510

Control of Pollution (Oil Storage) (England) Regulations 2001, SI 2001/2954

Control of Substances Hazardous to Health Regulations 2002, SI 2002/2677 [COSHH]

Environmental Impact Assessment (Scotland) Regulations 1999, SSI 1999/1

Environmental Protection (Duty of Care) Regulations 1991, SI 1991/2839 (as amended)

Groundwater Regulations 1998, SI 1998/2746

Groundwater Regulations (Northern Ireland) 1998, SR 1998/401

Hazardous Waste (England and Wales) Regulations 2005, SI 2005/894

List of Wastes (England) Regulations 2005, SI 2005/894 (as amended)

Management of Health and Safety at Work Regulations 1999, SI 1999/3242

Planning (Environmental Impact Assessment) Regulations (Northern Ireland) 1999, SR 1999/73

Special Waste Amendment (Scotland) Regulations 2004, SSI 2004/204

Special Waste Regulations (Northern Ireland) 1998, SR 1998/289

Town and Country Planning (Environmental Impact Assessment) (England and Wales) Regulations 1999, SI 1999/293

Town and Country Planning (General Development Procedure) Order 1995, SI 1995/419

Transport and Works (Model Clauses for Railways and Tramways) Order 1992, SI 1992/3270

Waste and Contaminated Land (Northern Ireland) Order 1997, SI 1997/2778 (N.I.19)

Water Environment (Controlled Activities) (Scotland) Regulations 2005, SSI 2005/348

Water Environment (Water Framework Directive) (England and Wales) Regulations 2003, SI 2003/3242

Water Environment (Water Framework Directive) Regulations (Northern Ireland) 2003, SR 2003/544

Water and Sewerage Services (Northern Ireland) Order 1973, SI 1973/70 (N.I.2), as amended

Water (Northern Ireland) Order 1999, SI 1999/662 (N.I.6)

STANDARDS

BS 1747-1 to 12:1969 to 1993. *Methods for measurement of air pollution*

BS 3247:1991. *Specification for salt for spreading on highways for winter maintenance*

BS 5667-6:2005. *Water quality. Sampling. Guidance on sampling of rivers and streams*

BS 5837:1991. *Trees in relation to construction. Recommendations*

BS 5930:1999. *Code of practice for site investigations*

BS 6068-6.11:1993. *Water quality. Sampling. Guidance on sampling of groundwaters*

BS 10175:2001. *Investigation of potentially contaminated sites. Code of practice*

ISO 14001:2004. *Environmental management systems. Requirements with guidance for use*

EA PUBLICATIONS

Pollution Prevention Guidelines (PPGs)

Published by Environment Agency, Bristol; SEPA, Edinburgh; EHS, Belfast

PPG1 *General guide to the prevention of pollution*, <www.sepa.org.uk/pdf/guidance/ppg/ppg01.pdf> (2001)

PPG2 *Above ground oil storage tanks*, <www.sepa.org.uk/pdf/guidance/ppg/ppg02.pdf> (2004)

PPG3 *Use and design of oil separators in surface water drainage systems*, <www.sepa.org.uk/pdf/guidance/ppg/ppg03.pdf> (2004)

PPG4 *Disposal of sewage where no mains drainage is available*, <www.sepa.org.uk/pdf/guidance/ppg/ppg04.pdf> (2004)

PPG5 *Works in, near or liable to affect watercourses*, <www.sepa.org.uk/pdf/guidance/ppg/ppg05.pdf> (2000)

PPG6 *Working at construction and demolition sites*, <www.sepa.org.uk/pdf/guidance/ppg/ppg06.pdf> (2003)

PPG7 *Refuelling facilities*, <www.sepa.org.uk/pdf/guidance/ppg/ppg07.pdf> (2004)

PPG8 *Safe storage and disposal of used oils*, <www.sepa.org.uk/pdf/guidance/ppg/ppg08.pdf> (2004)

PPG9 *Prevention of pollution by pesticides*, <www.sepa.org.uk/pdf/guidance/ppg/ppg09.pdf> (2000)

PPG11 *Preventing pollution on industrial sites*, <www.sepa.org.uk/pdf/guidance/ppg/ppg11.pdf> (2000)

PPG13 *High pressure water and steam cleaners*, <www.sepa.org.uk/pdf/guidance/ppg/ppg13.pdf> (2001)

PPG18 *Managing fire water and major spillages*, <www.sepa.org.uk/pdf/guidance/ppg/ppg18.pdf> (2000)

PPG20 *Dewatering of underground ducts and chambers*, <www.sepa.org.uk/pdf/guidance/ppg/ppg20.pdf> (2001)

PPG21 *Pollution incident response planning*, <www.sepa.org.uk/pdf/guidance/ppg/ppg21.pdf> (2004)

PPG22 *Dealing with spillages on highways*, <www.sepa.org.uk/pdf/guidance/ppg/ppg22.pdf> (2002)

PPG23 *Maintenance of structures over water*, <www.sepa.org.uk/pdf/guidance/ppg/ppg23.pdf> (2001)

PPG26 *Storage and handling of drums and intermediate bulk containers*, <www.sepa.org.uk/pdf/guidance/ppg/ppg26.pdf> (2004)

PPG27 *Installation, decommissioning and removal of underground storage tanks*, <www.sepa.org.uk/pdf/guidance/ppg/ppg27.pdf> (2002)

Technical guidelines

Published by EA, Bristol; SEPA, Edinburgh; EHS, Belfast; CIRIA, London

Masonry bunds for oil storage tanks, <www.sepa.org.uk/pdf/guidance/ppg/masonrybunds.pdf> (2000)

Concrete bunds for oil storage tanks, <www.sepa.org.uk/pdf/guidance/ppg/concretebunds.pdf> (2000)

WEBSITES

Association of Drainage Authorities <www.ada.org.uk>

Association of Inland Navigation Authorities, <www.aina.org.uk>

British Geological Survey, <www.bgs.ac.uk>

British Waterways, <www.britishwaterways.co.uk>

CEEQUAL, <www.ceequal.com>

CIRIA, <www.ciria.org>

Considerate Constructors Scheme, <www.considerateconstructorsscheme.org.uk>

Constructing Excellence, <www.constructingexcellence.org.uk/resourcecentre/kpizone>

Construction Confederation Environmental Forum, <www.thecc.org.uk>

Countryside Council for Wales, <www.ccw.gov.uk>

Defra groundwater guidance, <www.defra.gov.uk/environment/water/ground/guidance.htm>

Department of Trade and Industry, <www.dti.gov.uk/construction>

Environment Agency (England and Wales), <www.environment-agency.gov.uk>

Environment and Heritage Service (Northern Ireland), <www.ehsni.gov.uk>

Fisheries Research Services, <www.marlab.ac.uk>

Floodline: England and Wales, <www.environment-agency.gov.uk/floodline>

Floodline: N Ireland, <www.riversagencyni.gov.uk/rivers/floodemergency-whotocontact.htm>

Floodline: Scotland, <www.sepa.org.uk/flooding/index.html>

Highways Agency Research Compendium, <www.ha-research.co.uk>

International Erosion Control Association, <www.ieca.org>

Marine Consents and Environment Unit, <www.mceu.gov.uk/mceu_local/fepa/fepa-start.htm>

Met Office, <www.met-office.gov.uk>

Met Office MetBuild Direct service, <www.met-office.gov.uk/construction/mbdirect/index.html>

Multi-Agency Geographic Information for the Countryside (MAGIC), <www.magic.gov.uk>

National Water Archive, <www.nerc-wallingford.ac.uk/ih/nrfa/index.htm>

Natural England, <www.defra.gov.uk/ruraldelivery>

Netregs (environmental legislation), <www.netregs.co.uk>

Office of the Deputy Prime Minister, <www.odpm.gov.uk>

Office of Water Services, <www.ofwat.gov.uk>

Pipe Jacking Association, <www.pipejacking.org>

Rainwater reuse, <www.rainharvesting.co.uk>

Rivers Agency (Northern Ireland), <www.riversagencyni.gov.uk>

Scottish Environment Protection Agency, <www.sepa.org.uk>

Scottish Natural Heritage, <www.snh.org.uk>

Scottish Water, <www.scottishwater.gov.uk>

Sustainable drainage systems (CIRIA), <www.ciria.org.uk/suds>

Transport Analysis Guidance (DfT), <www.webtag.org.uk>

UK Society for Trenchless Technology, <www.ukstt.org.uk>

UK Spill Association, <www.ukspill.org>

Water Framework Directive guidance: England and Wales, <www.defra.gov.uk/environment/water/wfd/>

Water Framework Directive guidance: Northern Ireland, <www.ehsni.gov.uk/environment/waterManage/wfd/wfd.shtml>

Water Framework Directive guidance: Scotland, <www.sepa.org.uk/wfd>

Water safety, <www.healthandsafety.co.uk/water>

Water Service in Northern Ireland, <www.waterni.gov.uk >

Wheelwashing, <www.wheelwash.co.uk>